AF539407

Environmental Biology of Fishes

Environmental Biology of Fishes

Bhupad Joshi

RANDOM PUBLICATIONS
NEW DELHI - 110 002 (INDIA)

Environmental Biology of Fishes

ISBN 978-93-51112-85-3

Published in 2014 in India by
Reprint 2019

RANDOM PUBLICATIONS

4376-A/4B, Gali Murari Lal, Ansari Road
New Delhi-110 002
Phone: +9111-43580356, 23289044
E-mail: randomexports@gmail.com; sales@randompublications.com; info@randompublications.com

Type Setting by: Friends Media, Delhi-110089
Printed at : Mehra Printers, Delhi-110092

Preface

The origin of fishes dates back to over 480 million years. Fish evolved in fresh water; the chondrichthyes moved to the sea early in evolution, while the bony fishes went through most of their evolution in fresh water and spread to the seas at a much later period. Fish are the dominant free swimming animals of the seas. The structure of a fish body is designed for ease of movement. This ability to move about easily, without relying on water currents to carry them about, has enabled fishes to exploit most parts of the world's oceans, and this is reflected in an extraordinary variety of sizes and shapes. Fish are found in both fresh and salt water worldwide, and are a very important food source for many nations.

Commercial and subsistence fishers hunt fish in wild fisheries or farm them in ponds or in cages in the ocean. They are also caught by recreational fishers, kept as pets, raised by fishkeepers, and exhibited in public aquaria. The study of fish dates from the Upper Paleolithic Revolution (with the advent of 'high culture'). The science of ichthyology was developed in several interconnecting epochs, each with various significant advancements.

Successful survival in any environment depends upon an organism's ability to acquire information from its environment through its senses. Fish have many of the same senses that we have, they can see, smell, touch, feel, and taste, and they have developed some senses that we don't have, such as electroreception. Fish can sense light, chemicals, vibrations and electricity. In general, fishes are cold blooded. They derive their body heat from their environment and conform to its temperature. As water has a high heat capacity, it is able to easily suck any excess heat out of a fish and into the environment. They maintain a higher body temperature through the use of a specialized counter-current heat exchanger called a reta mirabile. These are dense capillary beds within the swimming muscle that run next to the veins leaving the muscles. Blood passes through the veins and arteries in a counter current (opposite) direction. The heat produced from the muscle contraction flows from the exiting veins into the incoming arteries and is recycled. The circulation of blood in fish is simple. The heart only has two chambers, in contrast to our heart which has four. This is because the fish heart only pumps blood in one direction. The blood enters the heart through a vein and exits through a vein on its way to the gills. In the gills, the blood picks up oxygen from the surrounding water and leaves the gills in arteries, which go to the body. The oxygen is used in the body and goes back to the heart.

Detailed coverage of environmental aspects relating to fish biology is a key feature of this book which will be of great use to students and individuals in fish biology, fisheries, aquaculture and environmental sciences.

I thank all members of my team who have helped in the preparation of the book. My special thanks go to "Random Publications" who have published the book.

—Bhupad Joshi

Contents

1

Introduction

A fish is any member of a paraphyletic group of organisms that consist of all gill-bearing aquatic craniate animals that lack limbs with digits. Included in this definition are the living hagfish, lampreys, and cartilaginous and bony fish, as well as various extinct related groups. Most fish are ectothermic ("cold-blooded"), allowing their body temperatures to vary as ambient temperatures change, though some of the large active swimmers like white shark and tuna can hold a higher core temperature. Fish are abundant in most bodies of water. They can be found in nearly all aquatic environments, from high mountain streams (e.g., char and gudgeon) to the abyssal and even hadal depths of the deepest oceans (e.g., gulpers and anglerfish). At 32,000 species, fish exhibit greater species diversity than any other group of vertebrates.

Fish are an important resource for humans worldwide, especially as food. Commercial and subsistence fishers hunt fish in wild fisheries or farm them in ponds or in cages in the ocean. They are also caught by recreational fishers, kept as pets, raised by fishkeepers, and exhibited in public aquaria. Fish have had a role in culture through the ages, serving as deities, religious symbols, and as the subjects of art, books and movies. Because the term "fish" is defined negatively, and excludes the tetrapods (i.e., the amphibians, reptiles, birds and mammals) which descend from within the same ancestry, it is paraphyletic, and is not considered a proper grouping in systematic biology. The traditional term pisces (also ichthyes) is considered a typological, but not a phylogenetic classification.

The earliest organisms that can be classified as fish were soft-bodied chordates that first appeared during the Cambrian period.

Although they lacked a true spine, they possessed notochords which allowed them to be more agile than their invertebrate counterparts. Fish would continue to evolve through the Paleozoic era, diversifying into a wide variety of forms. Many fish of the Paleozoic developed external armor that protected them from predators. The first fish with jaws appeared in the Silurian period, after which many (such as sharks) became formidable marine predators rather than just the prey of arthropods.

Evolution

Figure: *Fish do not represent a monophyletic group, and therefore the "evolution of fish" is not studied as a single event.*

Early fish from the fossil record are represented by a group of small, jawless, armored fish known as Ostracoderms. Jawless fish lineages are mostly extinct. An extant clade, the Lampreys may approximate ancient pre-jawed fish. The first jaws are found in Placodermi fossils. The diversity of jawed vertebrates may indicate the evolutionary advantage of a jawed mouth. It is unclear if the advantage of a hinged jaw is greater biting force, improved respiration, or a combination of factors. Fish may have evolved from a creature similar to a coral-like Sea squirt, whose larvae resemble primitive fish in important ways. The first ancestors of fish may have kept the larval form into adulthood (as some sea squirts do today), although perhaps the reverse is the case.

Anatomy

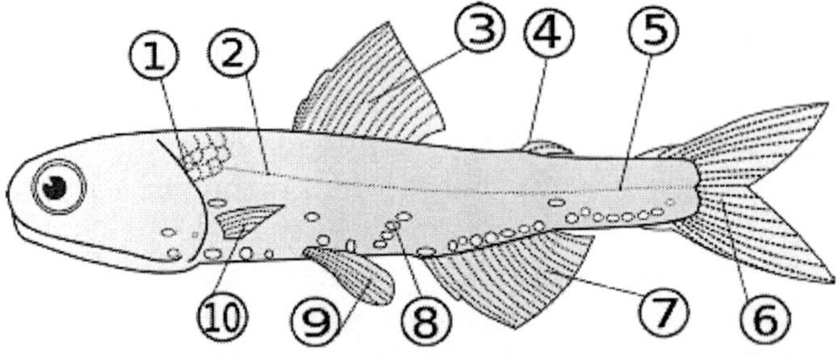

Figure: *The anatomy of* Lampanyctodes hectoris

(1) – operculum (gill cover), (2) – lateral line, (3) – dorsal fin, (4) – fat fin, (5) – caudal peduncle, (6) – caudal fin, (7) – anal fin, (8) – photophores, (9) – pelvic fins (paired), (10) – pectoral fins (paired)

Respiration

Most fish exchange gases using gills on either side of the pharynx. Gills consist of threadlike structures called filaments. Each filament contains a capillary network that provides a large surface area for exchanging oxygen and carbon dioxide. Fish exchange gases by pulling oxygen-rich water through their mouths and pumping it over their gills. In some fish, capillary blood flows in the opposite direction to the water, causing countercurrent exchange. The gills push the oxygen-poor water out through openings in the sides of the pharynx. Some fish, like sharks and lampreys, possess multiple gill openings. However, bony fish have a single gill opening on each side. This opening is hidden beneath a protective bony cover called an operculum.

Juvenile bichirs have external gills, a very primitive feature that they share with larval amphibians.

Fish from multiple groups can live out of the water for extended time periods. Amphibious fish such as the mudskipper can live and move about on land for up to several days, or live in stagnant or otherwise oxygen depleted water. Many such fish can breathe air via a variety of mechanisms. The skin of anguillid eels may absorb oxygen directly. The buccal cavity of the electric eel may breathe air. Catfish of the families Loricariidae, Callichthyidae, and Scoloplacidae absorb air through their digestive tracts. Lungfish, with the exception of the Australian lungfish, and bichirs have paired lungs similar to those of tetrapods and must surface to gulp fresh air through the mouth and pass spent air out through the gills. Gar and bowfin have a vascularized swim bladder that functions in the same way. Loaches, trahiras, and many catfish breathe by passing air through the gut. Mudskippers breathe by absorbing oxygen across the skin (similar to frogs). A number of fish have evolved so-called accessory breathing organs that extract oxygen from the air. Labyrinth fish (such as gouramis and bettas) have a labyrinth organ above the gills that performs this function. A few other fish have structures resembling labyrinth organs in form and function, most notably snakeheads, pikeheads, and the Clariidae catfish family.

Breathing air is primarily of use to fish that inhabit shallow, seasonally variable waters where the water's oxygen concentration may seasonally decline. Fish dependent solely on dissolved oxygen, such as perch and cichlids, quickly suffocate, while air-breathers survive for much longer, in some cases in water that is little more than wet mud. At the most extreme, some air-breathing fish are able to survive in damp burrows for weeks without water, entering a state of aestivation (summertime hibernation) until water returns.

Air breathing fish can be divided into obligate air breathers and facultative air breathers. Obligate air breathers, such as the African lungfish, *must* breathe air periodically or they suffocate. Facultative air breathers, such as the catfish *Hypostomus plecostomus*, only breathe air if they need to and will otherwise rely on their gills for oxygen. Most air breathing fish are facultative air breathers that avoid the energetic cost of rising to the surface and the fitness cost of exposure to surface predators.

Circulation

Fish have a closed-loop circulatory system. The heart pumps the blood in a single loop throughout the body. In most fish, the heart consists of four parts, including two chambers and an entrance and exit.

The first part is the sinus venosus, a thin-walled sac that collects blood from the fish's veins before allowing it to flow to the second part, the atrium, which is a large muscular chamber.

The atrium serves as a one-way antechamber, sends blood to the third part, ventricle.

The ventricle is another thick-walled, muscular chamber and it pumps the blood, first to the fourth part, bulbus arteriosus, a large tube, and then out of the heart.

The bulbus arteriosus connects to the aorta, through which blood flows to the gills for oxygenation.

Digestion

Jaws allow fish to eat a wide variety of food, including plants and other organisms. Fish ingest food through the mouth and break it down in the esophagus. In the stomach, food is further digested and, in many fish, processed in finger-shaped pouches called pyloric caeca, which secrete digestive enzymes and absorb nutrients.

Organs such as the liver and pancreas add enzymes and various chemicals as the food moves through the digestive tract. The intestine completes the process of digestion and nutrient absorption.

Scales

The scales of fish originate from the mesoderm (skin); they may be similar in structure to teeth.

Sensory and Nervous System

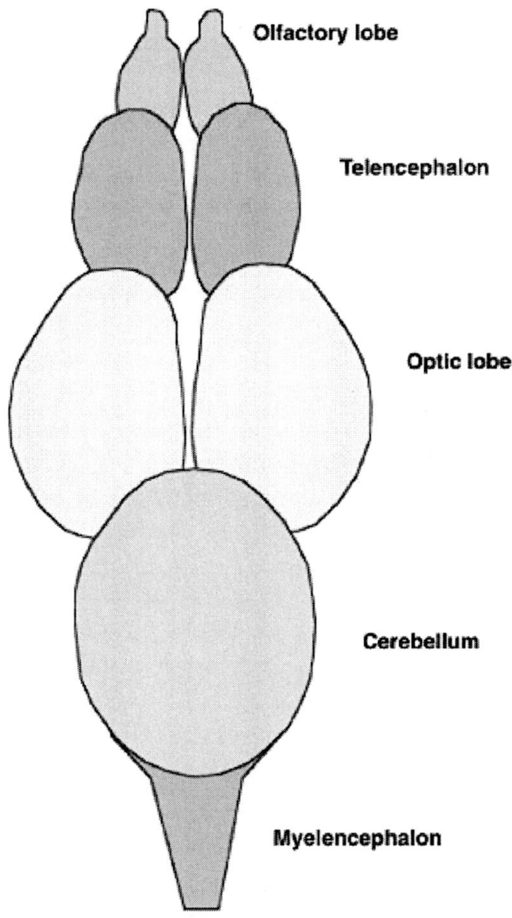

Figure: *Dorsal view of the brain of the rainbow trout*

Central Nervous System

Fish typically have quite small brains relative to body size compared with other vertebrates, typically one-fifteenth the brain mass of a similarly sized bird or mammal. However, some fish have relatively large brains, most notably mormyrids and sharks, which have brains about as massive relative to body weight as birds and marsupials.

Fish brains are divided into several regions. At the front are the olfactory lobes, a pair of structures that receive and process signals from the nostrils via the two olfactory nerves. The olfactory lobes are very large in fish that hunt primarily by smell, such as hagfish, sharks, and catfish. Behind the olfactory lobes is the two-lobed telencephalon, the structural equivalent to the cerebrum in higher vertebrates. In fish the telencephalon is concerned mostly with olfaction. Together these structures form the forebrain. Connecting the forebrain to the midbrain is the diencephalon (in the diagram, this structure is below the optic lobes and consequently not visible). The diencephalon performs functions associated with hormones and homeostasis. The pineal body lies just above the diencephalon. This structure detects light, maintains circadian rhythms, and controls colour changes.

The midbrain or mesencephalon contains the two optic lobes. These are very large in species that hunt by sight, such as rainbow trout and cichlids. The hindbrain or metencephalon is particularly involved in swimming and balance. The cerebellum is a single-lobed structure that is typically the biggest part of the brain. Hagfish and lampreys have relatively small cerebellae, while the mormyrid cerebellum is massive and apparently involved in their electrical sense. The brain stem or myelencephalon is the brain's posterior. As well as controlling some muscles and body organs, in bony fish at least, the brain stem governs respiration and osmoregulation.

Sense Organs

Most fish possess highly developed sense organs. Nearly all daylight fish have colour vision that is at least as good as a human's. Many fish also have chemoreceptors that are responsible for extraordinary senses of taste and smell. Although they have ears, many fish may not hear very well.

Most fish have sensitive receptors that form the lateral line system, which detects gentle currents and vibrations, and senses the motion of nearby fish and prey. Some fish, such as catfish and sharks, have organs that detect weak electric currents on the order of millivolt. Other fish, like the South American electric fishes Gymnotiformes, can produce weak electric currents, which they use in navigation and social communication. Fish orient themselves using landmarks and may use mental maps based on multiple landmarks or symbols. Fish behaviour in mazes reveals that they possess spatial memory and visual discrimination.

Capacity for Pain

Experiments done by William Tavolga provide evidence that fish have pain and fear responses. For instance, in Tavolga's experiments, toadfish grunted when electrically shocked and over time they came to grunt at the mere sight of an electrode.

In 2003, Scottish scientists at the University of Edinburgh and the Roslin Institute concluded that rainbow trout exhibit behaviours often associated with pain in other animals. Bee venom and acetic acid injected into the lips resulted in fish rocking their bodies and rubbing their lips along the sides and floors of their tanks, which the researchers concluded were attempts to relieve pain, similar to what mammals would do. Neurons fired in a pattern resembling human neuronal patterns.

Professor James D. Rose of the University of Wyoming claimed the study was flawed since it did not provide proof that fish possess "conscious awareness, particularly a kind of awareness that is meaningfully like ours". Rose argues that since fish brains are so different from human brains, fish are probably not conscious in the manner humans are, so that reactions similar to human reactions to pain instead have other causes. Rose had published a study a year earlier arguing that fish cannot feel pain because their brains lack a neocortex. However, animal behaviourist Temple Grandin argues that fish could still have consciousness without a neocortex because "different species can use different brain structures and systems to handle the same functions."

Animal welfare advocates raise concerns about the possible suffering of fish caused by angling. Some countries, such as Germany have banned specific types of fishing, and the British RSPCA now formally prosecutes individuals who are cruel to fish.

Homeothermy

Although most fish are exclusively ectothermic, there are exceptions.

Certain species of fish maintain elevated body temperatures. Endothermic teleosts (bony fish) are all in the suborder Scombroidei and include the billfishes, tunas, and one species of "primitive" mackerel (*Gasterochisma melampus*). All sharks in the family Lamnidae – shortfin mako, long fin mako, white, porbeagle, and salmon shark – are endothermic, and evidence suggests the trait exists in family Alopiidae (thresher sharks). The degree of endothermy varies from the billfish, which warm only their eyes and brain, to bluefin tuna and porbeagle sharks who maintain body temperatures elevated in excess of 20 °C above ambient water temperatures. Endothermy, though metabolically costly, is thought to provide advantages such as increased muscle strength, higher rates of central nervous system processing, and higher rates of digestion.

Reproductive System

Fish reproductive organs include testes and ovaries. In most species, gonads are paired organs of similar size, which can be partially or totally fused. There may also be a range of secondary organs that increase reproductive fitness.

In terms of spermatogonia distribution, the structure of teleosts testes has two types: in the most common, spermatogonia occur all along the seminiferous tubules, while in Atherinomorph fish they are confined to the distal portion of these structures. Fish can present cystic or semi-cystic spermatogenesis in relation to the release phase of germ cells in cysts to the seminiferous tubules lumen.

Figure: *Organs: 1. Liver, 2. Gas bladder, 3. Roe, 4. Pyloric caeca, 5. Stomach, 6. Intestine*

Fish ovaries may be of three types: gymnovarian, secondary gymnovarian or cystovarian. In the first type, the oocytes are released directly into the coelomic cavity and then enter the ostium, then through the oviduct and are eliminated. Secondary gymnovarian ovaries shed ova into the coelom from which they go directly into the oviduct. In the third type, the oocytes are conveyed to the exterior through the oviduct. Gymnovaries are the primitive condition found in lungfish, sturgeon, and bowfin. Cystovaries characterize most teleosts, where the ovary lumen has continuity with the oviduct. Secondary gymnovaries are found in salmonids and a few other teleosts.

Oogonia development in teleosts fish varies according to the group, and the determination of oogenesis dynamics allows the understanding of maturation and fertilization processes. Changes in the nucleus, ooplasm, and the surrounding layers characterize the oocyte maturation process. Postovulatory follicles are structures formed after oocyte release; they do not have endocrine function, present a wide irregular lumen, and are rapidly reabsorbed in a process involving the apoptosis of follicular cells. A degenerative process called follicular atresia reabsorbs vitellogenic oocytes not spawned. This process can also occur, but less frequently, in oocytes in other development stages.

Some fish are hermaphrodites, having both testes and ovaries either at different phases in their life cycle or, as in hamlets, have them simultaneously. Over 97% of all known fish are oviparous, that is, the eggs develop outside the mother's body. Examples of oviparous fish include salmon, goldfish, cichlids, tuna, and eels. In the majority of these species, fertilisation takes place outside the mother's body, with the male and female fish shedding their gametes into the surrounding water. However, a few oviparous fish practice internal fertilization, with the male using some sort of intromittent organ to deliver sperm into the genital opening of the female, most notably the oviparous sharks, such as the horn shark, and oviparous rays, such as skates. In these cases, the male is equipped with a pair of modified pelvic fins known as claspers.

Marine fish can produce high numbers of eggs which are often released into the open water column. The eggs have an average diameter of 1 millimetre (0.039 in).

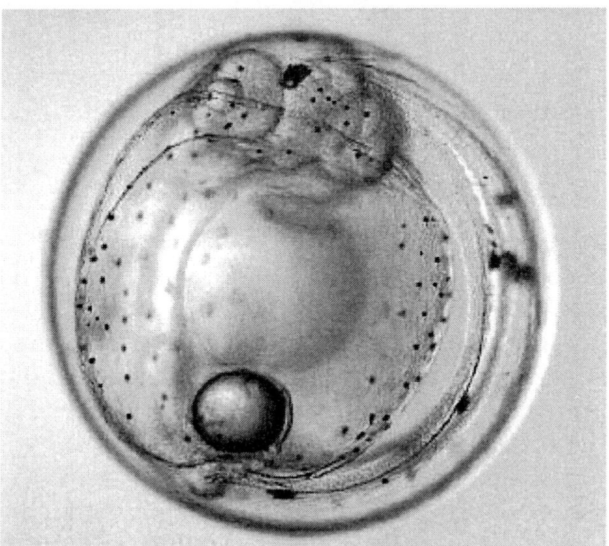

Figure: *An example of zooplankton*

The newly hatched young of oviparous fish are called larvae. They are usually poorly formed, carry a large yolk sac (for nourishment) and are very different in appearance from juvenile and adult specimens. The larval period in oviparous fish is relatively short (usually only several weeks), and larvae rapidly grow and change appearance and structure (a process termed metamorphosis) to become juveniles. During this transition larvae must switch from their yolk sac to feeding on zooplankton prey, a process which depends on typically inadequate zooplankton density, starving many larvae. In ovoviviparous fish the eggs develop inside the mother's body after internal fertilization but receive little or no nourishment directly from the mother, depending instead on the yolk. Each embryo develops in its own egg. Familiar examples of ovoviviparous fish include guppies, angel sharks, and coelacanths.

Some species of fish are viviparous. In such species the mother retains the eggs and nourishes the embryos. Typically, viviparous fish have a structure analogous to the placenta seen in mammals connecting the mother's blood supply with that of the embryo. Examples of viviparous fish include the surf-perches, splitfins, and lemon shark. Some viviparous fish exhibit oophagy, in which the developing embryos eat other eggs produced by the mother. This has been observed primarily among sharks, such as the shortfin mako and porbeagle, but is known for a few bony fish as well, such as the halfbeak *Nomorhamphus ebrardtii*. Intrauterine cannibalism is an even more unusual mode of vivipary, in which the largest embryos eat weaker and smaller siblings. This behaviour is also most commonly found among sharks, such as the grey nurse shark, but has also been reported for *Nomorhamphus ebrardtii*. Aquarists commonly refer to ovoviviparous and viviparous fish as livebearers.

Diseases

Like other animals, fish suffer from diseases and parasites. To prevent disease they have a variety of defences. *Non-specific* defences include the skin and scales, as well as the mucus layer secreted by the epidermis that traps and inhibits the growth of microorganisms. If pathogens breach these defences, fish can develop an inflammatory response that increases blood flow to the infected region and delivers white blood cells that attempt to destroy pathogens. Specific defences respond to particular pathogens recognised by the fish's body, i.e., an immune response. In recent years, vaccines have become widely used in aquaculture and also with ornamental fish, for example furunculosis vaccines in farmed salmon and koi herpes virus in koi.

Some species use cleaner fish to remove external parasites. The best known of these are the Bluestreak cleaner wrasses of the genus *Labroides* found on coral reefs in the Indian and Pacific Oceans. These small fish maintain so-called "cleaning stations" where other fish congregate and perform specific movements to attract the attention of the cleaners. Cleaning behaviours have been observed in a number of fish groups, including an interesting case between two cichlids of the same genus, *Etroplus maculatus*, the cleaner, and the much larger *Etroplus suratensis*.

Immune System

Immune organs vary by type of fish. In the jawless fish (lampreys and hagfish), true lymphoid organs are absent. These fish rely on regions of lymphoid tissue within other organs to produce immune cells. For example, erythrocytes, macrophages and plasma cells are produced in the anterior kidney (or pronephros) and some areas of the gut (where granulocytes mature.) They resemble primitive bone marrow in hagfish. Cartilaginous fish (sharks and rays) have a more advanced immune system. They have three specialised organs that are unique to chondrichthyes; the epigonal organs (lymphoid tissue similar to mammalian bone) that surround the gonads, the Leydig's organ within the walls of their esophagus, and a spiral valve in their intestine. These organs house typical immune cells (granulocytes, lymphocytes and plasma cells). They also possess an identifiable thymus and a well-developed spleen (their most important immune organ) where various lymphocytes, plasma cells and macrophages develop and are stored. Chondrostean fish (sturgeons, paddlefish and bichirs) possess a major site for the production of granulocytes within a mass that is associated with the meninges (membranes surrounding the central nervous system.) Their heart is frequently covered with tissue that contains lymphocytes, reticular cells and a small number of macrophages. The chondrostean kidney is an important hemopoietic organ; where erythrocytes, granulocytes, lymphocytes and macrophages develop.

Like chondrostean fish, the major immune tissues of bony fish (or teleostei) include the kidney (especially the anterior kidney), which houses many different immune cells. In addition, teleost fish possess a thymus, spleen and scattered immune areas within mucosal tissues (e.g. in the skin, gills, gut and gonads). Much like the mammalian immune system, teleost erythrocytes, neutrophils and granulocytes are believed to reside in the spleen whereas lymphocytes are the major cell type found in the thymus. In 2006, a lymphatic system similar to that in mammals was described in one species of teleost fish, the zebrafish. Although not confirmed as yet, this system presumably will be where naive (unstimulated) T cells accumulate while waiting to encounter an antigen.

Environmental Impact of Fishing

The environmental impact of fishing can be divided into issues that involve the availability of fish to be caught, such as overfishing, sustainable fisheries, and fisheries management; and issues that involve the impact of fishing on other elements of the environment, such as by-catch.

These conservation issues are part of marine conservation, and are addressed in fisheries science programs. There is a growing gap between how many fish are available to be caught and humanity's desire to catch them, a problem that gets worse as the world population grows. Similar to other environmental issues, there can be conflict between the fishermen who depend on fishing for their livelihoods and fishery scientists who realise that if future fish populations are to be sustainable then some fisheries must reduce or even close.

The journal *Science* published a four-year study in November 2006, which predicted that, at prevailing trends, the world would run out of wild-caught seafood in 2048. The scientists stated that the decline was a result of overfishing, pollution and other environmental factors that were reducing the population of fisheries at the same time as their ecosystems were being degraded. Yet again the analysis has met criticism as being fundamentally flawed, and many fishery management officials, industry representatives and scientists challenge the findings, although the debate continues. Many countries, such as Tonga, the United States, Australia and New Zealand, and international management bodies have taken steps to appropriately manage marine resources.

Ecological Disruption

Fishing may disrupt food webs by targeting specific, in-demand species. There might be too much fishing of prey species such as sardines and anchovies, thus reducing the food supply for the predators. It may also cause the increase of prey species when the target fishes are predator species such as salmon and tuna. Fisheries can reduce fish stocks that cetaceans rely on for food.

By-Catch

By-catch is the portion of the catch that is not the target species. These are either kept to be sold or discarded. In some instances the discarded portion is known as discards.

Possible Remedies

Many governments and intergovernmental bodies have implemented fisheries management policies designed to curb the environmental impact of fishing. Fishing conservation aims to control the human activities that may completely decrease a fish stock or washout an entire aquatic environment. These laws include the quotas on the total catch of particular species in a fishery, effort quotas (e.g., number of days at sea), the limits on the number of vessels allowed in specific areas, and the imposition of seasonal restrictions on fishing..

In 2008 a large scale study of fisheries that used individual transferable quotas and ones that didn't provided strong evidence that individual transferable quotas can help to prevent collapses and restore fisheries that appear to be in decline.

Fish farming has been proposed as a more sustainable alternative to traditional capture of wild fish. However, fish farming has been found to have negative impacts on nearby wild fish. Further, farming of predatory fish like salmon can rely on fish feed that is based on fish meal and oil from wild fish.

The environmental impact of recreational fishing may be alleviated to some extent by catch and release fishing.

Exotic Species

Introduction of non-native species has occurred in many habitats. One of the best studied examples is the introduction of Nile perch into Lake Victoria in the 1960s. Nile perch gradually exterminated the lake's 500 endemic cichlid species.

Some of them survive now in captive breeding programmes, but others are probably extinct. Carp, snakeheads, tilapia, European perch, brown trout, rainbow trout, and sea lampreys are other examples of fish that have caused problems by being introduced into alien environments.

Fishing Industry

The fishing industry includes any industry or activity concerned with taking, culturing, processing, preserving, storing, transporting, marketing or selling fish or fish products. It is defined by the FAO as including recreational, subsistence and commercial fishing, and the harvesting, processing, and marketing sectors. The commercial activity is aimed at the delivery of fish and other seafood products for human consumption or as input factors in other industrial processes. Directly or indirectly, the livelihood of over 500 million people in developing countries depends on fisheries and aquaculture.

Sectors

There are three principal industry sectors:

- The commercial sector: comprises enterprises and individuals associated with wild-catch or aquaculture resources and the various transformations of those resources into products for sale. It is also referred to as the "seafood industry", although non-food items such as pearls are included among its products.
- The traditional sector: comprises enterprises and individuals associated with fisheries resources from which aboriginal people derive products in accordance with their traditions.
- The recreational sector: comprises enterprises and individuals associated for the purpose of recreation, sport or sustenance with fisheries resources from which products are derived that are not for sale.

Commercial Sector

The commercial sector of the fishing industry comprises the following chain:

1. Commercial fishing and fish farming which produce the fish
2. Fish processing which produce the fish products
3. Marketing of the fish products

World Production

Fish are harvested by commercial fishing and aquaculture.

According to the Food and Agriculture Organisation (FAO), the world harvest in 2005 consisted of 93.3 million tonnes captured by commercial fishing in wild fisheries, plus 48.1 million tonnes produced by fish farms. In addition, 1.3 million tons of aquatic plants (seaweed etc.) were captured in wild fisheries and 14.8 million tons were produced by aquaculture. The number of individual fish caught in the wild has been estimated at 0.97-2.7 trillion per year (not counting fish farms or marine invertebrates).

Following is a table of the 2005 world fishing industry harvest in tonnes by capture and by aquaculture.

	Capture	***Aquaculture***	***Total***
Fish, crustaceans, molluscs, etc.	93,253,346	48,149,792	141,403,138
Aquatic plants	1,305,803	14,789,972	16,095,775
Total	94,559,149	62,939,764	157,498,913

This equates to about 24.4 kilograms a year for the average person on Earth.

Juvenile Fish

Fish go through various juvenile stages between birth and adulthood. They start as eggs which hatch into larva. The larva are not able to feed themselves, and carry a yolk-sac which provides their nutrition. Before the yolk-sac completely disappears, the tiny fish must become capable of feeding themselves. When they have developed to the point where are they capable of feeding themselves, the fish are called fry. When, in addition, they have developed scales and working fins, the transition to a juvenile fish is complete and it is called a fingerling. Fingerlings are typically about the size of fingers. The juvenile stage lasts until the fish is fully grown, sexually mature and interacting with other adult fish.

***Figure:** This young Chinook salmon, with scales and working fins, is a* fingerling

Growth Stages

Ichthyoplankton *(planktonic or drifting fish)* are the eggs and larvae of fish. They are usually found in the sunlit zone of the water column, less than 200 metres deep, sometimes called the epipelagic or photic zone. Ichthyoplankton are planktonic, meaning they cannot swim effectively under their own power, but must drift with ocean currents. Fish eggs cannot swim at all, and are unambiguously planktonic. Early stage larvae swim poorly, but later stage larvae swim better and cease to be planktonic as they grow into juveniles. Fish larvae are part of the zooplankton that eat smaller plankton, while fish eggs carry their own food supply. Both eggs and larvae are themselves eaten by larger animals.

According to Kendall et al. 1984 there are three main developmental stages of fish:

- Egg stage: Spawning to hatching. This stage is used instead of using an embryonic stage because there are aspects, such as those to do with the egg envelope, that are not just embryonic aspects.
- Larval stage: From hatching till all fin rays are present and the growth of fish scales has started (squamation). A key event is when the notochord associated with the tail fin on the ventral side of the spinal cord develops and becomes flexible. A transitional stage, the yolk-sac larval stage, lasts from hatching to the absorption of the yolk-sac.
- Juvenile stage: Starts when the transformation or metamorphosis from larva to juvenile is complete, that is, when the larva develops the features of a juvenile fish. These features are that all the fin rays are present and that scale growth is under way. The stage completes when the juvenile becomes adult, that is, when it becomes sexually mature or starts interacting with other adults.

This article is about the juvenile stage.

- Fry – refers to a recently hatched fish that has reached the stage where its yolk-sac has almost disappeared and its swim bladder is operational to the point where the fish can actively feed for itself.
- Fingerling – refers to a fish that has reached the stage where the fins can be extended and where scales have started developing throughout the body. It this stage, the fish is typically about the size of a finger.

Juvenile Salmon

Fry and fingerling are terms that can be applied to juvenile fish of most species. But some groups of fishes have juvenile development stages particular to the group. This section details the stages and the particular names used for juvenile salmon.

- *Sac fry* or *alevin* – The life cycle of salmon begins, and usually ends, in the backwaters of streams and rivers. These are the salmon spawning grounds, where salmon eggs are deposited, for safety, in the gravel. The salmon spawning grounds are also the salmon nurseries, providing a more protected environment than the ocean usually offers. After 2 to 6 months, the eggs hatch into tiny larvae, called *sac fry* or *alevin.* The alevin have a sac containing the remainder of the yolk, and they stay hidden in the gravel for a few days while they feed on the yolk.
- *Fry* – When the sac or yolk has almost gone, the baby fish must find food for themselves, so they leave the protection of the gravel and start feeding on plankton. At this point, the baby salmon are called *fry.*
- *Parr* – At the end of the summer, the fry develop into juvenile fish called *parr.* Parr feed on small invertebrates and are camouflaged with a pattern of spots and vertical bars. They remain in this stage for up to three years.
- *Smolt* – As they approach the time when they are ready to migrate out to the sea, the parr lose their camouflage bars and undergo a process of physiological changes, which allows them to survive a shift from freshwater to saltwater. At this point, the salmon are called *smolt.* Smolt spend time in the brackish waters of the river estuary while their body chemistry adjusts (osmoregulation) to the higher salt levels they will encounter in the ocean. Smolt also grow the silvery scales which visually confuse ocean predators.
- *Post-smolt* – When they have matured sufficiently in late spring, and are about 15 to 20 centimetres long, the smolt swim out of the rivers and into the sea. There they spend their first year as a *post-smolt.* Post-smolt form schools with other post-smolt, and set off to find deep-sea feeding grounds. They then spend up to four more years as adult ocean salmon while their full swimming and reproductive capacity develops.

Figure: *Salmon enter the ocean as* post-smolt *and mature into adult salmon. They gain most of their weight in the ocean*

Protection from Predators

Juvenile fish need protection from predators. Juvenile species, as with small species in general, can achieve some safety in numbers by schooling together. Juvenile coastal fish are drawn to turbid shallow waters and to mangrove structures, where they have better protection from predators. As the fish grow, their foraging ability increases and their vulnerability to predators decreases, and they tend to shift from mangroves to mudflats. In the open sea juvenile species often aggregate around floating objects such as jellyfish and *Sargassum* seaweed. This can significantly increase their survival rates.

As Food

Juvenile fish are marketed as food.

- *Whitebait* is a marketing term for the fry of fish, typically between 25 and 50 millimetres long. Such juvenile fish often travel together in schools along the coast, and move into estuaries and sometimes up rivers where they can be easily caught with fine meshed fishing nets. *Whitebaiting* is the activity of catching whitebait. Whitebait are tender and edible, and can be regarded as a delicacy. The entire fish is eaten including head, fins and gut. Some species make better eating than others, and the particular species that are marketed as "whitebait" varies in different parts of the world. As whitebait consists of fry of many important food species (such as herring, sprat, sardines, mackerel, bass and many others) it is not an ecologicially viable foodstuff and in several countries strict controls on harvesting exist.

- *Elvers* are young eels. Traditionally, fishermen consumed elvers as a cheap dish, but environmental changes have reduced eel populations. Similar to whitebait, they are now considered a delicacy and are priced at up to 1000 euro per kilogram. Spain has eel dishes, called "angulas", which are baby eels usually served sauteed in olive oil, garlic and a chili pepper. Elvers are now very expensive. A small serving of angulas can cost equivalent of $US100, and there are imitation angulas which can be purchased cheaply.

Commercial Fishing

Commercial fishing is the activity of catching fish and other seafood for commercial profit, mostly from wild fisheries. It provides a large quantity of food to many countries around the world, but those who practice it as an industry must often pursue fish far into the ocean under adverse conditions. Large-scale commercial fishing is also known as industrial fishing. This profession has gained in popularity with the development of shows such as *Deadliest Catch*, *Swords*, and *Wicked Tuna*. The major fishing industries are not only owned by major corporations but by small families as well. The industry has had to adapt through the years in order to keep earning a profit. A study taken on some small family-owned commercial fishing companies showed that they adapted to continue to earn a living but not necessarily make a large profit. It is the adaptability of the fishermen and their methods that cause some concern for fishery managers and researchers; they say that for those reasons, the sustainability of the marine ecosystems could be in danger of being ruined.

Commercial fishermen harvest a wide variety of animals, ranging from tuna, cod, carp, and salmon to shrimp, krill, lobster, clams, squid, and crab, in various fisheries for these species.

There are large and important fisheries worldwide for various species of fish, mollusks, crustaceans, and echinoderms. However, a very small number of species support the majority of the world's fisheries. Some of these species are herring, cod, anchovy, tuna, flounder, mullet, squid, shrimp, salmon, crab, lobster, oyster and scallops. All except these last four provided a worldwide catch of well over a million tonnes in 1999, with herring and sardines together providing a catch of over 22 million metric tons in 1999. Many other species are fished in smaller numbers.

The industry, in 2006, also managed to generate over 185 billion dollars in sales and also provide over two million jobs, according to an economic report released by NOAA's Fisheries Service. Commercial fishing may offer an abundance of jobs, but the pay varies from boat to boat, season to season.

Crab fisherman Cade Smith was quoted in an article by *Business Week* as saying, "There was always a top boat where the crew members raked in $50,000 during the three- to five-day king crab season—or $100,000 for the longer snow crab season". That may be true, but there are also the boats who don't do well; Smith said later in the same article that his worst season left him with a loss of 500 dollars.

A 2009 paper in *Science* estimates, for the first time, the total world fish biomass as somewhere between 0.8 and 2.0 billion tonnes.

Fish Farming

Aquaculture is the cultivation of aquatic organisms. Unlike fishing, aquaculture, also known as aquafarming, is the cultivation of aquatic populations under controlled conditions. Mariculture refers to aquaculture practiced in marine environments. Particular kinds of aquaculture include algaculture (the production of kelp/seaweed and other algae); fish farming; shrimp farming, shellfish farming, and the growing of cultured pearls. Fish farming involves raising fish commercially in tanks or enclosed pools, usually for food. Fish species raised by fish farms include carp, salmon, tilapia, catfish and cod. Increasing demands on wild fisheries by commercial fishing operations have caused widespread overfishing. Fish farming offers an alternative solution to the increasing market demand for fish and fish protein.

Mariculture

Mariculture is a specialised branch of aquaculture involving the cultivation of marine organisms for food and other products in the open ocean, an enclosed section of the ocean, or in tanks, ponds or raceways which are filled with seawater. An example of the latter is the farming of marine fish, including finfish and shellfish e.g.prawns, or oysters and seaweed in saltwater ponds. Non-food products produced by mariculture include: fish meal, nutrient agar, jewellery (e.g. cultured pearls), and cosmetics.

Environmental Effects of Fish Farming

Commonly identified environmental impacts from marine farms are:

1. Wastes from cage cultures;
2. Farm escapees and invasives;

3. Genetic pollution and disease and parasite transfer;
4. Habitat modification.

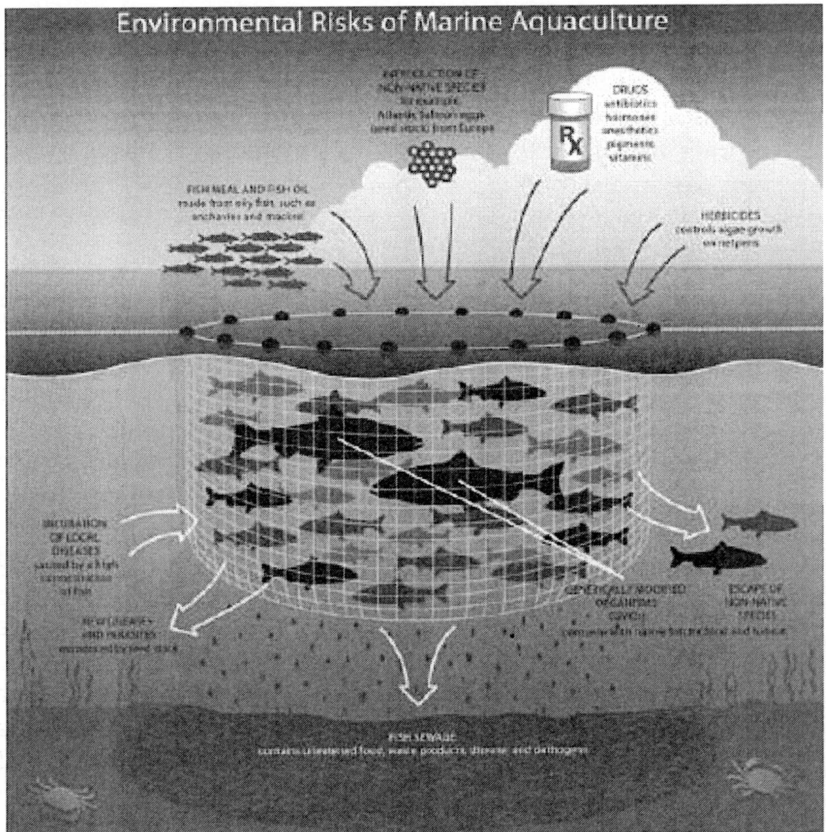

Figure: *Mariculture has rapidly expanded over the last two decades due to new technology, improvements in formulated feeds, greater biological understanding of farmed species, increased water quality within closed farm systems, greater demand for seafood products, site expansion and government interest. As a consequence, mariculture has been subject to some controversy regarding its social and environmental impacts.*

As with most farming practices, the degree of environmental impact depends on the size of the farm, the cultured species, stock density, type of feed, hydrography of the site, and husbandry methods. The adjacent diagram connects these causes and effects.

Genetic Pollution and Disease and Parasite Transfer

One of the primary concerns with mariculture is the potential for disease and parasite transfer. Farmed stocks are often selectively bred to increase disease and parasite resistance, as well as improving growth rates and quality of products. As a consequence, the genetic diversity within reared stocks decreases with every generation - meaning they can potentially reduce the genetic diversity within wild populations if they escape into those wild populations. Such genetic pollution from escaped aquaculture stock can reduce the wild population's ability to adjust to the changing natural environment. Also, maricultured species can harbour diseases and parasites (e.g., lice) which can be introduced to wild populations upon their escape. An example of this is the parasitic sea lice on wild and farmed Atlantic salmon in Canada.

Also, non-indigenous species which are farmed may have resistance to, or carry, particular diseases (which they picked up in their native habitats) which could be spread through wild populations if they escape into those wild populations. Such 'new' diseases would be devastating for those wild populations because they would have no immunity to them.

Habitat Modification

With the exception of benthic habitats directly beneath marine farms, most mariculture causes minimal destruction to habitats. However, the destruction of mangrove forests from the farming of shrimps is of concern. Globally, shrimp farming activity is a small contributor to the destruction of mangrove forests; however, locally it can be devastating. Mangrove forests provide rich matrices which support a great deal of biodiversity – predominately juvenile fish and crustaceans. Furthermore, they act as buffering systems whereby they reduce coastal erosion, and improve water quality for in situ animals by processing material and 'filtering' sediments.

Fish Processing

The term fish processing refers to the processes associated with fish and fish products between the time fish are caught or harvested, and the time the final product is delivered to the customer. Although the term refers specifically to fish, in practice it is extended to cover any aquatic organisms harvested for commercial purposes, whether caught in wild fisheries or harvested from aquaculture or fish farming.

Larger fish processing companies often operate their own fishing fleets or farming operations. The products of the fish industry are usually sold to grocery chains or to intermediaries. Fish are highly perishable. A central concern of fish processing is to prevent fish from deteriorating, and this remains an underlying concern during other processing operations. Fish processing can be subdivided into fish handling, which is the preliminary processing of raw fish, and the manufacture of fish products. Another natural subdivision is into primary processing involved in the filleting and freezing of fresh fish for onward distribution to fresh fish retail and catering outlets, and the secondary processing that produces chilled, frozen and canned products for the retail and catering trades.

There is evidence humans have been processing fish since the early Holocene. These days, fish processing is undertaken by artisan fishermen, on board fishing or fish processing vessels, and at fish processing plants. Fish is a highly perishable food which needs proper handling and preservation if it is to have a long shelf life and retain a desirable quality and nutritional value. The central concern of fish processing is to prevent fish from deteriorating. The most obvious method for preserving the quality of fish is to keep them alive until they are ready for cooking and eating. For thousands of years, China achieved this through the aquaculture of carp. Other methods used to preserve fish and fish products include

- the control of temperature using ice, refrigeration or freezing
- the control of water activity by drying, salting, smoking or freeze-drying
- the physical control of microbial loads through microwave heating or ionizing irradiation
- the chemical control of microbial loads by adding acids
- oxygen deprivation, such as vacuum packing.

Usually more than one of these methods is used. When chilled or frozen fish or fish products are transported by road, rail, sea or air, the cold chain must be maintained. This requires insulated containers or transport vehicles and adequate refrigeration. Modern shipping containers can combine refrigeration with a controlled atmosphere.

Fish processing is also concerned with proper waste management and with adding value to fish products. There is an increasing demand for ready to eat fish products, or products that don't need much preparation.

Preservation

Preservation techniques are needed to prevent fish spoilage and lengthen shelf life. They are designed to inhibit the activity of spoilage bacteria and the metabolic changes that result in the loss of fish quality. Spoilage bacteria are the specific bacteria that produce the unpleasant odours and flavours associated with spoiled fish. Fish normally host many bacteria that are not spoilage bacteria, and most of the bacteria present on spoiled fish played no role in the spoilage. To flourish, bacteria need the right temperature, sufficient water and oxygen, and surroundings that are not too acidic. Preservation techniques work by interrupting one or more of these needs. Preservation techniques can be classified as follows.

Control of Temperature

Figure: *Ice preserves fish and extends shelf life by lowering the temperature*

If the temperature is decreased, the metabolic activity in the fish from microbial or autolytic processes can be reduced or stopped. This is achieved by refrigeration where the temperature is dropped to about 0 °C, or freezing where the temperature is dropped below -18°C. On fishing vessels, the fish are refrigerated mechanically by circulating cold air or by packing the fish in boxes with ice. Forage fish, which are often caught in large numbers, are usually chilled with refrigerated or chilled seawater. Once chilled or frozen, the fish need further cooling to maintain the low temperature. There are key issues with fish cold store design and management, such as how large and energy efficient they are, and the way they are insulated and palletized.

An effective method of preserving the freshness of fish is to chill with ice by distributing ice uniformly around the fish. It is a safe cooling method that keeps the fish moist and in an easily stored form suitable for transport. It has become widely used since the development of mechanical refrigeration, which makes ice easy and cheap to produce. Ice is produced in various shapes; crushed ice and ice flakes, plates, tubes and blocks are commonly used to cool fish. Particularly effective is slurry ice, made from micro crystals of ice formed and suspended within a solution of water and a freezing point depressant, such as common salt.

A more recent development is pumpable ice technology. Pumpable ice flows like water, and because it is homogeneous, it cools fish faster than fresh water solid ice methods and eliminates freeze burns. It complies with HACCP and ISO food safety and public health standards, and uses less energy than conventional fresh water solid ice technologies.

2

Aquatic Predation

Aquatic predation presents a special difficulty as compared to predation on land, because the density of water is about the same as that of the prey, so that the prey tends to be pushed away. This problem was first identified by R.N. Alexander. As a result, underwater predators, especially bony fish, have evolved a number of specialised feeding mechanisms, known as filter feeding, ram feeding, suction feeding, protrusion, and pivot feeding. Most underwater predators combine more than one of these basic principles. For example a typical generalised predator, such as the cod, combines suction with some amount of protrusion and pivot feeding.

Suction Feeding

Suction feeding is a method of ingesting a prey item in fluids by sucking the prey into the predator's mouth. This is typically accomplished by the predator expanding the volume of its oral cavity and/or throat, resulting in a pressure difference between the inside of the mouth and the outside environment. When the mouth is opened, the pressure difference causes water to flow into the predator's mouth, carrying the prey item in with the fluid flow. Bony fish have evolved ingenious mechanisms, so-called mechanical linkages for rapid and forceful expansion of the buccal cavity.

Ram Feeding

Ram feeding, also known as lunge feeding, is a method of feeding underwater in which the predator moves forward with its mouth open, engulfing the prey along with the water surrounding it. During ram feeding, the prey remains fixed in space, and the predator moves its jaws past the prey to capture it. The motion of the head may induce a bow wave in the fluid which pushes the prey away from the jaws, but this can be avoided by allowing water to flow through the jaw. This can be accomplished by means of an expandable throat, as in snapping turtles and baleen whales, or by allowing water to flow out through the gills, as in sharks and herring. A number of species have evolved narrow snouts, as in gar fish and water snakes.

Herrings often hunt copepods. If they encounter copepods schooling in high concentrations, the herrings switch to ram feeding. They swim with their mouth wide open and their opercula fully expanded. Every several feet, they close and clean their gill rakers for a few milliseconds (filter feeding). The fish all open their mouths and opercula wide at the same time (the red gills are visible in the photo above—click to enlarge). The fish swim in a grid where the distance between them is the same as the jump length of the copepods.

Suspension Feeding

In suspension feeding, the water flow is primarily external or if the particles themselves move with respect to the ambient water, such as in sea lilies.

Bait Ball

A bait ball, or baitball, occurs when small fish swarm in a tightly packed spherical formation about a common centre. It is a last ditch defensive measure adopted by small schooling fish when they are threatened by predators. Small schooling fish are eaten by many types of predators, and for this reason they are called bait fish or forage fish. For example, sardines group together when they are threatened. This instinctual behaviour is a defence mechanism, as lone individuals are more likely to be eaten than large groups. Sardine bait balls can be 10–20 metres in diametre and extend to a depth of 10 metres. The bait balls are short lived and seldom last longer than 10 minutes.

However, bait balls are also conspicuous, and when schooling fish form a bait ball they can draw the attention of many other predators. As a response to the defensive capabilities of schooling fish, some predators have developed sophisticated countermeasures. These countermeasures can be spectacularly successful, and can seriously undermine the defensive value of forming bait balls.

Formation and Dissolution

The process that leads to the formation of a bait ball typically starts when predators locate a fish school deep below the surface. The predators make rushes and use various scare tactics to force the fish school to the surface, herding it at the same time into a compact volume. The alarmed fish, trapped against the surface above and surrounded all about, abandon their coordinated schooling movements and become chaotic. Their graceful and disciplined schooling strategies of uniform spacing and polarity degrade into frenetic attempts by each fish to save itself. In this way, a dense bait ball forms as each fish scrambles to get away from the surface of the ball and hide in the interior. The symmetry of this centripetal action forms a sphere, the shape with the minimum surface area for a given volume. This exposes the fewest number of fish on the surface to the predators. The movement, sound and smell can attract more predators, including different predator species, until there is a carousel of them, each species using their own characteristic predatory strategies. Fish that break loose are singled out and eaten. A frenzy can develop as predators compete, the water reddening with blood as shredded flesh and scales drift to the depths. As the bait ball reduces in size and number, it becomes progressively easier for the predators to target the remaining survivors.

Predator Strategies

Predators have devised various countermeasures to disrupt the defensive shoaling and schooling manoeuvres of forage fish. Often this involves charging the school or bait ball at high speed.

Some whales *lunge feed* on bait balls. Lunge feeding is an extreme feeding method, in which the whale accelerates from below a bait ball to a high velocity and then opens its mouth to a large gape angle. This generates the water pressure required to expand its mouth and engulf and filter a huge amount of water and fish. Lunge feeding by the huge rorqual whales is said to be the largest biomechanical event on Earth.

Swordfish charge at high speeds through forage fish schools, slashing with their swords to kill or stun prey. They then turn and return to consume their catch. Thresher sharks use their long tails to stun shoaling fishes. Spinner sharks charge vertically through schools, spinning on their axis with their mouths open and snapping all around. The shark's momentum at the end of these spiralling runs often carries it into the air.

Gannets plummet from heights of 30 metres (100 feet), plunging through the water leaving vapour-like trails behind like fighter planes. They enter the water at speeds up to 86 kilometres per hour (53 mph) and descend to depths of 34 metres (111 feet). Gannets have air sacs under their skin in their faces and chests which act like bubble-wrap, cushioning the impact with water.

Cleaner Fish

Cleaner fish are fish that provide a service to other fish species by removing dead skin and ectoparasites. This is an example of mutualism, an ecological interaction that benefits both parties involved. A wide variety of fishes have been observed to display cleaning behaviors including wrasses, cichlids, catfish, and gobies, as well as by a number of different species of cleaner shrimp. There is also at least one predatory mimic, the sabre-toothed blenny, that mimics cleaner fish but in fact feeds on healthy scales and mucous.

Diversity of Cleaner Fish

Marine Fishes: The best known cleaner fish are the cleaner wrasses of the genus *Labroides* found on coral reefs in the Indian Ocean and Pacific Ocean. These small fish maintain so-called cleaning stations where other fish, known as hosts, will congregate and perform specific movements to attract the attention of the cleaner fish. Remarkably, these small cleaner fish will safely clean large predatory fish that would otherwise eat small fish such as these. Cleaner wrasses appear to get almost all their nutrition through this cleaning service, and when maintained in aquaria rarely survive for long because they cannot obtain enough to eat.

Cleaning behaviors have been observed in a number of other fish groups. Neon gobies of the genera *Gobiosoma* and *Elacatinus* provide a cleaning service similar to the cleaner wrasses, though this time on reefs in the Western Atlantic, providing a good example of convergent evolution. Unlike the cleaner wrasses, they also eat a variety of small animals as well being cleaner fish, and generally do well in aquaria. However, the Caribbean cleaning goby (*Elacatinus evelynae*) will gladly eat scales and mucus from the host when the ectoparasites it normally feeds on are scarce, making the relationship somewhat less than mutually beneficial. The symbiosis does not break down because the abundance of these parasites varies significantly seasonally and spacially, and the overall benefit to the larger fish outweighs any cheating on the part of the smaller.

Brackish Water Fishes

An interesting example of a cleaning symbiosis has been observed between two brackish water cichlids of the genus *Etroplus* from South Asia. The small species *Etroplus maculatus* is the cleaner fish, and the much larger *Etroplus suratensis* is the host that receives the cleaning service.

Freshwater Fishes

Cleaning is notably less common in freshwater habitats than in marine habitats. One of the few examples of cleaning is juvenile Striped Raphael catfish cleaning the piscivorous *Hoplias* cf. *malabaricus*.

Mimicry

The sabre-toothed blenny *Aspidontus taeniatus* is a blenny that mimics the ritualised dance the cleaner wrasse makes when passing fish swim by. Instead of providing a useful cleaning service, however, it bites off pieces of healthy skin and scales from the host before darting away to safety.

Filter Feeder

Filter feeders (a sub-group of suspension feeders) are animals that feed by straining suspended matter and food particles from water, typically by passing the water over a specialised filtering structure. Some animals that use this method of feeding are clams, krill, sponges, some fish and sharks, and baleen whales. Some birds, such as flamingos, are also filter feeders. Filter feeders can play an important role in clarifying water.

Baleen Whales

The baleen whales, also called whalebone whales or great whales, form the Mysticeti, one of two suborders of the Cetacea (whales, dolphins, and porpoises). Baleen whales are characterised by having baleen plates for filtering food from water, rather than having teeth. This distinguishes them from the other suborder of cetaceans, the toothed whales or Odontoceti. The suborder contains four families and fourteen species. The scientific name derives from the Greek word mystidos, which means "unknowable".

Bivalves

Bivalves are aquatic molluscs which have two-part shells. Typically both shells (or valves) are symmetrical along the hinge line. The class has 30,000 species, including scallops, clams, oysters and mussels. Most bivalves are filter feeders (although some have taken up scavenging and predation), extracting organic matter from the sea in which they live. Nephridia, the shell fish version of kidneys, remove the waste material. Buried bivalves feed by extending a siphon to the surface.

As an example, oysters draw water in over their gills through the beating of cilia. Suspended food (phytoplankton, zooplankton, algae and other water-borne nutrients and particles) are trapped in the mucus of a gill, and from there are transported to the mouth, where they are eaten, digested and expelled as feces or pseudofeces. Each oyster filters up to five litres of water per hour. Scientists believe that the Chesapeake Bay's once-flourishing oyster population historically filtered the estuary's entire water volume of excess nutrients every three or four days. Today that process would take almost a year, and sediment, nutrients, and algae can cause problems in local waters. Oysters filter these pollutants, and either eat them or shape them into small packets that are deposited on the bottom where they are harmless.

Sponges

Sponges have no true circulatory system; instead, they create a water current which is used for circulation. Dissolved gases are brought to cells and enter the cells via simple diffusion. Metabolic wastes are also transferred to the water through diffusion. Sponges pump remarkable amounts of water. *Leuconia*, for example, is a small leuconoid sponge about 10 cm tall and 1 cm in diametre. It is estimated that water enters through more than 80,000 incurrent canals at a speed of 6 cm per minute. However, because *Leuconia* has more than 2 million flagellated chambers whose combined diametre is much greater than that of the canals, water flow through chambers slows to 3.6 cm per hour. Such a flow rate allows easy food capture by the collar cells. All water is expelled through a single osculum at a velocity of about 8.5 cm/second: a jet force capable of carrying waste products some distance away from the sponge.

Flamingos

Flamingos filter-feed on brine shrimp. Their oddly-shaped beaks are specially adapted to separate mud and silt from the food they eat, and are uniquely used upside-down. The filtering of food items is assisted by hairy structures called lamellae which line the mandibles, and the large rough-surfaced tongue.

Other Filter Feeders

Other examples of filter-feeding organisms include:

- Brachiopoda (a phylum within the Eumetazoa subkingdom of animals, commonly referred to as lamp shells)

- Branchiopoda (a class within the Arthropod phylum of animals. Branchiopods are primitive crustaceans found mostly in freshwater)
- Bryozoa (a phylum within the Eumetazoa subkingdom of animals, commonly referred to as moss animals or sea mats)
- Canalipalpata (an order of Polychaete ringed worms, commonly referred to as bristle-footed annelids)
- Ctenophora (a phylum within the Eumetazoa subkingdom of animals, commonly referred to as comb jellies)
- Echiura (an order of Polychaete ringed worms, commonly referred to as spoon worms)
- Entoprocta (a phylum within the Eumetazoa subkingdom of animals, commonly referred to as goblet worms)
- Lobodon carcinophagus (a subphylum of the Chordata phylum of animals, commonly referred to as Crabeater seals)
- Holothuroidea (a class within the Echinodermata phylum of animals, commonly referred to as sea cucumbers)
- Phoronida (a phylum within the Eumetazoa subkingdom of animals, commonly referred to as horseshoe worms)
- Sipuncula (a phylum within the Eumetazoa subkingdom of animals, commonly referred to as peanut worms)
- Urochordata (a subphylum of the Chordata phylum of animals, commonly referred to as sea squirts).

Forage Fish

Forage fish, also called prey fish or bait fish, are small fish which are preyed on by larger predators for food. Predators include other larger fish, seabirds and marine mammals. Typical ocean forage fish feed near the base of the food chain on plankton, often by filter feeding. They include the fishes of the family Clupeidae (herrings, shad, sardines, hilsa, menhaden and sprats), as well as anchovies, capelin and halfbeaks.

Forage fish compensate for their small size by forming schools. Some swim in synchronised grids with their mouths open so they can efficiently filter plankton. These schools can become immense shoals which move along coastlines and migrate across open oceans. The shoals are concentrated fuel resources for the great marine predators. The predators are keenly focused on the shoals, acutely aware of their numbers and whereabouts, and make migrations themselves that can span thousands of miles to connect, or stay connected, with them. The ocean primary producers, mainly contained in plankton, produce food energy from the sun and are the raw fuel for the ocean food webs. Forage fish transfer this energy by eating the plankton and becoming food themselves for the top predators.

In this way, forage fish occupy the central positions in ocean and lake food webs.

In recent times, many of the worlds great predator fisheries have collapsed. To compensate, the fishing industry is removing huge amounts of forage fish from the oceans, using factory ships with sophisticated sonar and spotting planes. Most of the catch is fed to farmed animals. Fisheries scientists are expressing concern that this will result in further collapses of the predator fish that depend on them.

Diet

Forage fish feed on plankton. When they are eaten by larger predators, they transfer this energy from the bottom of the food chain to the top and in this way are the central link between trophic levels.

Forage fish are usually filter feeders, meaning that they feed by straining suspended matter and food particles from water. They usually travel in large, slow moving, tightly packed schools with their mouths open. They are typically omnivorous. Their diet is usually based primarily on zooplankton, although, since they are omnivorous, they also take in some phytoplankton.

Young forage fish, such as herring, mostly feed on phytoplankton and as they mature they start to consume larger organisms. Older herrings feed on zooplankton, tiny animals that are found in oceanic surface waters, and fish larvae and fry (recently-hatched fish). Copepods and other tiny crustaceans are common zooplankton eaten by forage fish. During daylight, many forage fish stay in the safety of deep water, feeding at the surface only at night when there is less chance of predation. They swim with their mouths open, filtering plankton from the water as it passes through their gills.

Ocean halfbeaks are omnivores which feed on algae, plankton, marine plants like seagrass, invertebrates like pteropods and crustaceans and smaller fishes. Some tropical species feed on animals during the day and plants at night, while others alternate summer carnivory with winter herbivoury. They are in turn eaten by billfish, mackerel, and sharks.

Predators

Forage fish are the food that sustains larger predators above them in the ocean food chain. The superabundance they present in their schools make them ideal food sources for top predator fish such as tuna, striped bass, cod, salmon, barracuda and swordfish, as well as sharks, whales, dolphins, porpoises, seals, sea lions, and seabirds.

Hunting Copepods

Copepods are a group of small crustaceans found in ocean and freshwater habitats. Many species are planktonic (drifting in the ocean water), while others are benthic (living on the sea floor). Copepods are typically one millimetre (0.04 in) to two millimetres (0.08 in) long, with a teardrop shaped body. Like other crustaceans they have an armoured exoskeleton, but they are so small that this armour, and the entire body, is usually transparent.

Copepods are usually the dominant zooplankton. Some scientists say they form the largest animal biomass on the planet. The other contender is the Antarctic krill. But copepods are smaller than krill, with faster growth rates, and they are more evenly distributed throughout the oceans. This means copepods almost certainly contribute more secondary production to the world's oceans than krill, and perhaps more than all other groups of marine organisms together. They are a major item on the forage fish menu.

Copepods are very alert and evasive. They have large antennae. When they spread their antennae they can sense the pressure wave from an approaching fish and jump with great speed over a few centimetres.

Herrings are pelagic feeders. Their prey consists of a wide spectrum of phytoplankton and zooplankton, amongst which copepods are the dominant prey. Young herring usually capture small copepods by hunting them individually— they approach them from below. The (half speed) video loop at the left shows a juvenile herring feeding on copepods. In the middle of the image a copepod escapes successfully to the left. The opercula (hard bony flaps covering the gills) are spread wide open to compensate the pressure wave which would alert the copepod to trigger a jump.

If prey concentrations reach very high levels, the herrings adopt a method called "ram feeding". They swim with their mouth wide open and their opercula fully expanded. Every several feet, they close and clean their gill rakers for a few milliseconds (filter feeding). In the photo on the right, herring ram feed on a school of copepods. The fish all open their mouths and opercula wide at the same time (the red gills are visible—click to enlarge). The fish swim in a grid where the distance between them is the same as the jump length of their prey, as indicated in the animation below.

In the animation, juvenile herring hunt the copepods in synchronisation: The copepods sense with their antennae the pressure-wave of an approaching herring and react with a fast escape jump. The length of the jump is fairly constant. The fish align themselves in a grid with this characteristic jump length. A copepod can dart about 80 times before it tires out. After a jump, it takes it 60 milliseconds to spread its antennae again, and this time delay becomes its undoing, as the almost endless stream of herrings allows a herring to eventually snap the copepod. A single juvenile herring could never catch a large copepod.

Migrations

Forage fish often make great migrations between their spawning, feeding and nursery grounds. Schools of a particular stock usually travel in a triangle between these grounds. For example, one stock of herrings have their spawning ground in southern Norway, their feeding ground in Iceland, and their nursery ground in northern Norway. Wide triangular journeys such as these may be important because forage fish, when feeding, cannot distinguish their own offspring.

Fertile feeding grounds for forage fish are provided by ocean upwellings. Oceanic gyres are large-scale ocean currents caused by the Coriolis effect. Wind-driven surface currents interact with these gyres and the underwater topography, such as seamounts and the edge of continental shelves, to produce downwellings and upwellings. These can transport nutrients which plankton thrive on. The result can be rich feeding grounds attractive to the plankton feeding forage fish. In turn, the forage fish themselves become a feeding ground for larger predator fish. Most upwellings are coastal, and many of them support some of the most productive fisheries in the world. Regions of notable upwelling include coastal Peru, Chile, Arabian Sea, western South Africa, eastern New Zealand and the California coast.

Capelin are a forage fish of the smelt family found in the Atlantic and Arctic oceans. In summer, they graze on dense swarms of plankton at the edge of the ice shelf. Larger capelin also eat krill and other crustaceans. The capelin move inshore in large schools to spawn and migrate in spring and summer to feed in plankton rich areas between Iceland, Greenland, and Jan Mayen. The migration is affected by ocean currents. Around Iceland maturing capelin make large northward feeding migrations in spring and summer. The return migration takes place in September to November. The spawning migration starts north of Iceland in December or January.

The diagram on the right shows the main spawning grounds and larval drift routes. Capelin on the way to feeding grounds is coloured green, capelin on the way back is blue, and the breeding grounds are red. In a paper published in 2009, researchers from Iceland recount their application of an interacting particle model to the capelin stock around Iceland, successfully predicting the spawning migration route for 2008.

Contemporary

Traditional commercial fisheries were directed towards high value ocean predators such as cod, rockfish and tuna, rather than humble forage fish. As technologies developed, fisheries became so effective at locating and catching predator fish that many of the stocks collapsed. The industry compensated by turning to species lower in the food chain. In former times, forage fish were more difficult to fish profitably, and were a small part of the global marine fisheries. But modern industrial fishing technologies have enabled the removal of increasing quantities. Industrial-scale forage fish fisheries need large scale landings of fish to return profits. They are dominated by a small number of corporate fishing and processing companies.

Forage fish populations are very vulnerable when faced with modern fishing equipment. They swim near the surface in compacted schools, so they are relatively easy to locate at the surface with sophisticated electronic fishfinders and from above with spotter planes. Once located, they are scooped out of the water using highly efficient nets, such as purse seines, which remove most of the school.

Spawning patterns in forage fish are highly predictable. Some fisheries use knowledge of these patterns to harvest the forage species as they come together to spawn, removing the fish before they have actually spawned. Fishing during spawning periods or at other times when forage fish amass in large numbers can also be a blow to predators.

Many predators, such as whales, tuna and sharks, have evolved to migrate long distances to specific sites for feeding and breeding. Their survival hinges on their finding these forage schools at their feeding grounds. The great ocean predators find that, no matter how they are adapted for speed, size, endurance or stealth, they are on the losing side when faced with the machinery of contemporary industrial fishing.

Altogether, forage fish account for 37 percent (31.5 million tonnes) of all fish taken from the world's oceans each year. However, because there are fewer species of forage fish compared to predator fish, forage species fisheries are the largest in the world. Seven of the top ten fisheries target forage fish. The total world catch of herrings, sardines and anchovies alone in 2005 was 22.4 million tonnes, 24 percent of the total world catch.

The Peruvian anchoveta fishery is now the biggest in the world (10.7 million tonnes in 2004), while the Alaskan pollock fishery in the Bering Sea is the largest single species fishery in the world (3 million tonnes). The Alaskan pollock is said to be the largest remaining single species source of palatable fish in the world. However, the biomass of pollock has declined in recent years, perhaps spelling trouble for both the Bering Sea ecosystem and the commercial fishery it supports.

Acoustic surveys by NOAA indicate that the 2008 pollock population is almost 50 percent lower than last year's survey levels. Some scientists think this decline in Alaska pollock could repeat the collapse experienced by Atlantic cod, which could have negative consequences for the entire Bering Sea ecosystem. Salmon, halibut, endangered Steller sea lions, fur seals, and humpback whales eat pollock and depend on healthy populations to sustain themselves.

In Lakes and Rivers

Forage fish also inhabit freshwater habitats, such as lakes and rivers, where they serve as food for larger freshwater predators. Usually smaller than 15 centimetres (6 in) in length, these small bait fish make up most of the fish found in lakes and rivers. The minnow family alone, consisting of minnows, chubs, shiners and daces, consists of more than fifty species. Other freshwater forage fish include suckers, killifish, shad, bony fish as well as fish of the sunfish family, excluding black basses and crappie, and smaller species of the carp family. There are also anadromous forage fish, such as eulachon.

Within any fresh or saltwater ecosystem, there will always be both desirable and undesirable fishes, and this varies from country to country, and often from region to region within a country. Sport fishermen divide freshwater predators of forage fish into those:

- which have a good fighting ability and are good to eat, called sport (or game) fish.
- the other less desirable fish, called rough fish in North America and coarse fish in Britain

Rough or coarse fish usually refers to fish that are not commonly eaten, not sought after for sporting reasons, or have become invasive species reducing the populations of desirable fish. They compete for forage fish with the more popular sport fish. They are often regarded as a nuisance, and are not usually protected by game laws. Forage fish generally are not considered rough or coarse fish because of their usefulness as bait.

The term *rough fish* is used by U.S. State agencies and anglers to describe undesirable predator fish. In North America, anglers fish for salmon, trout, bass, pike, catfish, walleye and muskellunge. The smallest fish are called panfish, because they can fit in a standard cooking pan. Some examples are crappies, rock bass, perch, bluegill and sunfish.

The term *coarse fish* originated in the United Kingdom in the early 19th century. Prior to that time, recreational fishing was the sport of the gentry, who angled for trout and salmon which they called "game fish". Fish other than game fish were disdained as "coarse fish". These days, "game fish" refers to Salmonids (other than grayling) — that is, salmon, trout and char. Coarse fish are made up mostly of the larger species of Cyprinids (carp, roach, bream) as well as pike, catfish, gar and lamprey. Coarse fish are no longer disdained; indeed, fishing for coarse fish has become a popular pastime.

Bait and Feeder Fish

Forage fish are sometimes referred to as *bait fish* or *feeder fish.* Bait fish is a term used particularly by recreational fishermen, although commercial fisherman also catch fish to bait longlines and traps. Forage fish is a fisheries term, and is used in the context of fisheries. Bait fish, by contrast, are fish that are caught by humans to use as bait for other fish.

The terms overlap in the sense that most bait fish are also forage fish, and vice versa. Feeder fish is a term used particularly in the context of fish aquariums. It refers essentially to the same concept as forage fish, small fish that are eaten by larger fish, but the term is adapted to the particular requirements of working with fish in aquariums.

Fish Migration

Many types of fish migrate on a regular basis, on time scales ranging from daily to annually or longer, and over distances ranging from a few metres to thousands of kilometres. Fish usually migrate because of diet or reproductive needs, although in some cases the reason for migration remains unknown.

Other Examples

Some of the best-known anadromous fish are the six species of Pacific salmon, which are Chinook (King), Coho (Silver), Sockeye (Red), Chum (Dog), Pink (Humpback), and Cherry. The salmon hatch in small freshwater streams. From there they migrate to the sea to mature, living there for two to six years. When mature, the salmon return to the same streams where they were hatched to spawn. Salmon are capable of going hundreds of kilometres upriver, and humans must install fish ladders in dams to enable the salmon to get past. Other examples of anadromous fishes are sea trout, three-spined stickleback, and shad.

The most remarkable catadromous fishes are freshwater eels of genus *Anguilla*, whose larvae drift from swawning grounds in the Sargasso sea, sometimes for months or years, before entering freshwater river and streams as glass eels or elvers.

An example of a euryhaline species is the Bull shark, which lives in Lake Nicaragua of Central America and the Zambezi River of Africa. Both these habitats are fresh water, yet Bull sharks will also migrate to and from the ocean. Specifically, Lake Nicaragua Bull sharks migrate to the Atlantic Ocean and Zambezi Bull sharks migrate to the Indian Ocean.

Diel vertical migration is a common behaviour; many marine species move to the surface at night to feed, then return to the depths during daytime.

A number of large marine fishes, such as the tuna, migrate north and south annually, following temperature variations in the ocean. These are of great importance to fisheries.

Freshwater fish migrations are usually shorter, typically from lake to stream or vice versa, for spawning purposes. However, potamodromous migrations of Colorado pikeminnow of the Colorado River system can be extensive. Migrations to natal spawning grounds easily be 100 km, with maximum distances of 300 km reported from radiotagging studies.

Historic Exploitation

Since prehistoric times humans have exploited certain anadromous fishes during their migrations into freshwater streams, when they are more vulnerable to capture. Societies dating to the Millingstone Horizon are known which exploited the anadromous fishery of Morro Creek and other Pacific coast estuaries. In Nevada the Paiute tribe has harvested migrating Lahontan cutthroat trout along the Truckee River since prehistoric times. This fishing practice continues to current times, and the U.S. Environmental Protection Agencyý has supported research to assure the water quality in the Truckee can support suitable populations of the Lahontan cutthroat trout.

Modelling Fish Migration

In a paper published in 2009, researchers from Iceland recount their application of an interacting particle model to the capelin stock around Iceland, successfully predicting the spawning migration route for 2008.

Shoaling and Schooling

In biology, any group of fish that stay together for social reasons are said to be shoaling (pronounced /ÈƒoŠljK/), and if, in addition, the group is swimming in the same direction in a coordinated manner, they are said to be schooling (pronounced /ÈskuÐljK/). In common usage, the terms are sometimes used rather loosely. About one quarter of fishes shoal all their lives, and about one half of fishes shoal for part of their lives.

Fish derive many benefits from shoaling behaviour including defence against predators (through better predator detection and by diluting the chance of individual capture), enhanced foraging success, and higher success in finding a mate. It is also likely that fish benefit from shoal membership through increased hydrodynamic efficiency. Fish use many traits to choose shoalmates. Generally they prefer larger shoals, shoalmates of their own species, shoalmates similar in size and appearance to themselves, healthy fish, and kin (when recognised).

The "oddity effect" posits that any shoal member that stands out in appearance will be preferentially targeted by predators. This may explain why fish prefer to shoal with individuals that resemble themselves. The oddity effect would thus tend to homogenize shoals.

An aggregation of fish is the general term for any collection of fish that have gathered together in some locality. Fish aggregations can be structured or unstructured. An unstructured aggregation might be a group of mixed species and sizes that have gathered randomly near some local resource, such as food or nesting sites.

If, in addition, the aggregation comes together in an interactive, social way, they are said to be shoaling. Although shoaling fish can relate to each other in a loose way, with each fish swimming and foraging somewhat independently, they are nonetheless aware of the other members of the group as shown by the way they adjust behaviour such as swimming, so as to remains close to the other fish in the group. Shoaling groups can include fish of disparate sizes and can including mixed-species subgroups.

If, as a further addition, the shoal becomes more tightly organised, with the fish synchronising their swimming so they all move at the same speed and in the same direction, then the fish are said to be schooling. Schooling fish are usually of the same species and the same age/size. Fish schools move with the individual members precisely spaced from each other. The schools undertake complicated manoeuvres, as though the schools have minds of their own.

Shoaling is a special case of aggregating, and schooling is a special case of shoaling. While schooling and shoaling mean different things within biology, they are often treated as synonyms by non-specialists, with speakers of British English tending to use "shoaling" to describe any grouping of fish, while speakers of American English tend to use "schooling" just as loosely. The intricacies of schooling are far from fully understood, especially the swimming and feeding energetics. Many hypotheses to explain the function of schooling have been suggested, such as better orientation, synchronised hunting, predator confusion and reduced risk of being found. Schooling also has disadvantages, such as excretion buildup in the breathing media and oxygen and food depletion. The way the fish array in the school probably gives energy saving advantages, though this is controversial.

Fish can be obligate or facultative shoalers. Obligate shoalers, such as tunas, herrings and anchovy, spend all of their time shoaling or schooling, and become agitated if separated from the group. Facultative shoalers, such as Atlantic cod, saiths and some carangids, shoal only some of the time, perhaps for reproductive purposes.

Shoaling fish can shift into a disciplined and coordinated school, then shift back to an amorphous shoal within seconds. Such shifts are triggered by changes of activity from feeding, resting, travelling or avoiding predators.

When schooling fish stop to feed, they break ranks and become shoals. Shoals are more vulnerable to predator attack. The shape a shoal or school takes depends on the type of fish and what the fish are doing. Schools that are travelling can form long thin lines, or squares or ovals or amoeboid shapes. Fast moving schools usually form a wedge shape, while shoals that are feeding tend to become circular.

Forage fish are small fish which are preyed on by larger predators for food. Predators include other larger fish, seabirds and marine mammals. Typical ocean forage fish are small, filter feeding fish such as herring, anchovies and menhaden. Forage fish compensate for their small size by forming schools. Some swim in synchronised grids with their mouths open so they can efficiently filter feed on plankton. These schools can become huge, moving along coastlines and migrating across open oceans. The shoals are concentrated fuel resources for the great marine predators.

These sometimes immense gatherings fuel the ocean food web. Most forage fish are pelagic fish, which means they form their schools in open water, and not on or near the bottom (demersal fish). Forage fish are short-lived, and go mostly unnoticed by humans, apart from an occasional support role in a documentary about a great ocean predator. The predators are keenly focused on the shoals, acutely aware of their numbers and whereabouts, and make migrations themselves, often in schools of their own, that can span thousands of miles to connect with, or stay connected with them.

Herring are among the more spectacular schooling fish. They aggregate together in huge numbers. The largest schools are often formed during migrations by merging with smaller schools. "Chains" of schools one hundred kilometres long have been observed of mullet migrating in the Caspian Sea. Radakov estimated herring schools in the North Atlantic can occupy up to 4.8 cubic kilometres with fish densities between 0.5 and 1.0 fish/cubic metre.

That's about three billion fish in one school. These schools move along coastlines and traverse the open oceans. Herring schools in general have very precise arrangements which allow the school to maintain relatively constant cruising speeds. Herrings have excellent hearing, and their schools react very fast to a predator. The herrings keep a certain distance from a moving scuba diver or cruising predator like a killer whale, forming a vacuole which looks like a doughnut from a spotter plane.

Many species of large predatory fish also school, including many highly migratory fish, such as tuna and some ocean going sharks. Cetaceans such as dolphins, porpoises and whales, operate in organised social groups called pods.

"Shoaling behaviour is generally described as a trade-off between the anti-predator benefits of living in groups and the costs of increased foraging competition." Landa (1998) argues that the cumulative advantages of shoaling, as elaborated below, are strong selective inducements for fish to join shoals. Parrish et al. (2002) argue similarly that schooling is a classic example of emergence, where there are properties that are possessed by the school but not by the individual fish. Emergent properties give an evolutionary advantage to members of the school which non members do not receive.

Hydrodynamic Efficiency

This theory states that groups of fish may save energy when swimming together, much in the way that bicyclists may draft one another in a peloton. Geese flying in a Vee formation are also thought to save energy by flying in the updraft of the wingtip vortex generated by the previous animal in the formation. Increased efficiencies in swimming in groups have been proposed for schools of fish and Antarctic krill.

It would seem reasonable to think that the regular spacing and size uniformity of fish in schools would result in hydrodynamic efficiencies. However, experiments in the laboratory have failed to find any gains from the hydrodynamic lift created by the neighbours of a fish within a school, though it is still thought that efficiency gains do occur in the wild. Landa (1998) argues that the leader of a school constantly changes, because while being in the body of a school gives a hydrodynamic advantage, being the leader means you are the first to the food.

Mapping the Formation of Schools

In 2009, building on recent advances in acoustic imaging, a group of MIT researchers observed for "the first time the formation and subsequent migration of a huge shoal of fish." The results provide the first field confirmation of general theories about how large groups behave, from locust swarms to bird flocks.

The researchers imaged spawning Atlantic herring off Georges Bank. They found that the fish come together from deeper water in the evening, shoaling in a disordered way. A chain reaction triggers when the population density reaches a critical value, like an audience wave travelling around a sport stadium. A rapid transition then occurs, and the fish become highly polarised and synchronised in the manner of schooling fish. After the transition, the schools start migrating, extending up to 40 kilometres (25 mi) across the ocean, to shallow parts of the bank. There they spawn during the night. In the morning, the fish school back to deeper water again and then disband. Small groups of leaders were also discovered that significantly influenced much larger groups.

Making Decisions

Fish schools are faced with decisions they must make if they are to remain together. For example, a decision might be which direction to swim when confronted by a predator, which areas to stop and forage, or when and where to migrate. How are these decisions made? Do more experienced 'leaders' exert more influence than other group members, or does the group make a decision by consensus? A recent investigation showed that small groups of fish used consensus decision-making when deciding which fish model to follow. The fish did this by a simple quorum rule such that individuals watched the decisions of others before making their own decisions. This technique generally resulted in the 'correct' decision but occasionally cascaded into the 'incorrect' decision. In addition, as the group size increased, the fish made more accurate decisions in following the more attractive fish model. Consensus decision-making, a form of collective intelligence, thus effectively uses information from multiple sources to generally reach the correct conclusion.

Other open questions of shoaling behaviour include identifying which individuals are responsible for the direction of shoal movement. In the case of migratory movement, most members of a shoal seem to know where they are going. Observations on the foraging behaviour of captive golden shiner (a kind of minnow) found they formed shoals which were led by a small number of experienced individuals who knew when and where food was available. In a herd of sheep the same lead sheep are always in the front row. But in a school of migrating fish, if the school changes direction, the fish previously at the flank become the lead fish.

One puzzling aspect of shoal selection is how a fish can choose to join a shoal of animals similar to themselves, given that it cannot know its own appearance. Experiments with zebrafish have shown that shoal preference is a learned ability, not innate. A zebrafish tends to associate with shoals that resemble shoals in which it was reared (that is, a form of imprinting).

Commercial Fishing

The schooling behaviour of fish is exploited on an industrial scale by the commercial fishing industry. Huge purse seiner vessels use spotter planes to locate schooling fish, such as tuna, cod, mackerel and forage fish. They can capture huge schools by rapidly encircling them with purse seine nets with the help of fast auxiliary boats and sophisticated sonar, which can track the shape of the shoal.

Light Senses

Vision is used by animals to determine the layout of their surroundings, and thus this sense is particularly important for locomotion. In animals with eyes that have good resolution, vision can be used to identify objects from their geometric appearance; however, this requires a sophisticated brain of the kind found in vertebrates, cephalopod mollusks such as octopus, and higher arthropods, such as bees and jumping spiders. All vision, or photoreception, relies on photo receptors that contain a special light-detecting molecule known as rhodopsin. Rhodopsin detects electromagnetic radiation—light with wavelengths in the range 400–700 nanometres (1 nm = 10m). There are some animals that can detect infrared radiation (wavelengths greater than 700 nm); for example, some snakes use infrared radiation to locate warm-blooded prey, and certain beetles can use it to sense forest fires. However, animals that detect wavelengths in the infrared do this with receptors that sense heat or mechanical expansion, rather than with photoreceptors.

The rhodopsin molecule of photoreceptors consists of a protein called opsin that straddles the cell membrane with seven helices. These form a structure with a central cavity that contains a chromophore group, which in humans is called retinal—the aldehyde of vitamin A. When retinal absorbs a photon of light, it changes its configuration (from the bent 11-*cis* form to the straight all-*trans* form), setting off a series of molecular reactions that lead, within a few milliseconds, to a change in the flow of ions through the cell membrane.

Many invertebrates have the capacity to see and analyze polarised light. Polarisation arises from atmospheric scattering and reflection at smooth surfaces such as water. In polarised light all the photons have their electrical fields vibrating in the same plane; this can be detected by photoreceptors if the molecules are appropriately aligned. The projecting microvillus structure of invertebrate receptors makes this possible. Many insects use Polarisation to work out the Sun's direction when the sky is overcast, and others use it to detect water surfaces.

The optical systems of eyes break down light according to its direction of origin and thus form images that can be used for navigation and pattern recognition. There are about 10 ways of forming images, including pinholes, lenses, and mirrors. Of these, the single-chambered "camera-type" eyes of vertebrates and cephalopods have the best resolution. The human eye can resolve stripes spaced 1 minute of arc ($^1/_{60}$ of 1°) apart; this is many times better than the compound eye of a bee, which can resolve objects spaced about 2.8°–5.4° apart.

Mechanical Senses

There are a great many varieties of mechanical receptors in animals, but best known are the receptors that mediate touch, the variety of hair cell receptors in vertebrates that mediate hearing (the acoustico-lateralis system), and the muscle spindle proprioceptors that monitor the state of muscle contraction. The basic mechanism by which a stimulus is converted to an electrical signal in cells is known as transduction. There are six types of touch receptors in human skin, including free nerve endings, hair follicle receptors, Meissner corpuscles, Merkel endings, Ruffini endings, and Pacinian corpuscles. The first three, free nerve endings, hair follicle receptors, and Meissner corpuscles, respond to superficial light touch; the next two, Merkel endings and Ruffini endings, to touch pressure; and the last one, Pacinian corpuscles, to vibration. Pacinian corpuscles are built in a way that gives them a fast response and quick recovery. They contain a central nerve fibre surrounded by onion like layers of connective tissue that behave like a shock absorber, transmitting fast events but damping out slow changes. The fibre, which on its own is capable of sustained firing, only responds to rapid events with one or two action potentials.

In all vertebrates there is a type of mechanically sensitive cell known as a hair cell. The outer surface of these cells contains an array of tiny hairlike processes, including a kinocilium (not present in mammals), which has a typical internal fibre skeleton, and stereocilia, which do not have fibre skeletons. Stereocilia decrease in size with distance from the kinocilium and are functionally polarised. When the stereocilia are bent toward the kinocilium, the hair cell is excited, and the nerve fibre that contacts the cell fires action potentials. In contrast, bending the hairs away from the kinocilium inhibits firing.

- Hair cells have many uses. In fishes the cells are part of the lateral line system, a series of canals in the skin that are open to the surrounding water and that are used to monitor water currents caused by the fish itself and by other fish. The canals are equipped at intervals with clusters of hair cells, each with a jellylike cap known as a cupula. The cupula is displaced by water movement, thus bending the hairs beneath it, resulting in activity in the nerve. In the inner ear of higher vertebrates there are three variants of this basic design, responsible for detecting the direction of gravity, angular rotation, and sound waves. In the utricle and saccule of the inner ear there are patches of hair cells known as maculae. Within each maculae, the stereocilia are embedded in a gelatinous mass known as the otolithic membrane, which contains small stonelike calcium carbonate particles called otoconia.

In mammals the ear consists of the outer sound-collecting pinna; the middle ear, which contains ossicles that function to match the mechanics of sound in air to sound in water; and the inner ear, which contains the cochlea. The cochlea is a complex coiled structure. It consists of a long membrane, known as the basilar membrane, which is tuned in such a way that high tones vibrate the region near the base and low tones vibrate the region near the apex. Sitting on the basilarmembrane is the organ of Corti, an array of hair cells with stereocilia that contact a gelatinous membrane called the tectorial membrane. Sound entering the inner ear stimulates different regions of the basilar membrane, depending on sound frequency. Hair cells in the stimulated regions are excited by the resulting shearing action between the stereocilia and the tectorialmembrane. There are two kinds of hair cells in the organ of Corti. The inner hair cells are sensory, and the nerves extending from them send acoustic information to the brain. In contrast, the outer hair cells are motile and have a role in amplifying and modifying the movement of the basilar membrane.

The human ear is sensitive to sounds ranging in frequency from 20 hertz to 20 kilohertz. Below about 1 kilohertz, frequency is signaled by the actual frequency of action potentials in the auditory nerve; above this frequency, however, it is the region of the basilarmembrane that vibrates most that specifies frequency.

A special type of mechanical receptor is found in muscles. These mechanoreceptors are known as muscle spindles and consist of the stretch-sensitive endings of one or more neurons attached to a region near the centre of a modified muscle fibre. This fibre has its own innervation, independent of the innervation of the main muscle. The neurons projecting from the muscle spindle respond to lengthening of the muscle. However, by activating the muscle attached to the receptor, the spindle can be stretched or relaxed independently, thereby setting the range over which it will respond to changes in length of the main muscle. This double innervation provides the brain with a very flexible way of activating muscles and of monitoring load-induced stretch.

A number of other minor senses are probably best thought of as mechanical senses. Pain often originates from mechanical action, although, where tissue damage results, the stimulus may involve chemical action as well. In some animals, including bees and pigeons, there is evidence that a magnetic sense is involved in navigation. In these animals magnetite grains have been found in suitable physiological sensory reception locations. It has been proposed that movement of these grains may act as either a locational or a directional stimulus.

3

Internal Structure: Vascular and Neural

Vascular

Blood enters the gills through afferent branchial arteries (ABAs), which receive the entire cardiac output via the ventral aorta. The blood that flows through an ABA feeds the two hemibranchs of an arch, where the blood is oxygenated at the lamellae of the filaments. Oxygenated blood from the filaments of an arch is collected by an efferent branchial artery (EBA), which directs blood to the dorsal aorta for systemic distribution. Upon closer examination of blood flow through the filaments, two distinct, yet interconnected, circulatory systems are apparent: the arterio-arterial and arteriovenous vasculature. Because of the large variations in the anatomy of these systems among species (especially in the arteriovenous vasculature), it is difficult to make generalisations. Therefore, the description that follows is not intended to be an all-encompassing description of the gill circulatory pathways, but aims to introduce some of the more important anatomical features of the vasculature that relate to functional aspects of the physiological processes discussed later in this review. For more thorough accounts of gill vascular anatomy.

Arterio-arterial Vasculature

The arterio-arterial vasculature is often referred to as the respiratory pathway, because it is responsible for the exchange of gases between a fish's blood and its environment. Because of the general similarities in this circulatory system among elasmobranchs, lampreys, and teleosts, these groups will be described together; the respiratory pathway of hagfishes will be described separately. In the arterio-arterial system, blood from an ABA feeds filaments on the hemibranchs of an arch via afferent filamental arteries (AFAs), which travel along the length of a filament. In elasmobranchs and lampreys, the AFA regularly feeds the corpus cavernosum (CC) or cavernous body, which is an extensive network of interconnected vascular sinuses in the afferent portion of the filament, e.g., the spiny dogfish; little skate, *Raja erinacea*; lesser-spotted dogfish, *Scyliorhinus canicula*; Endeavour dogfish, *Centrophorus scalpratus*; sparsely spotted stingaree, *Urolophus paucimaculatus*; Western shovelnose stingaree, *Trygonoptera mucosa*; and Arctic lamprey, *Lampetra japonica*.

Possible functions of the CC are as a hydrostatic support device for the gill filament, a blood pressure regulator, and a site of erythrocyte phagocytosis.

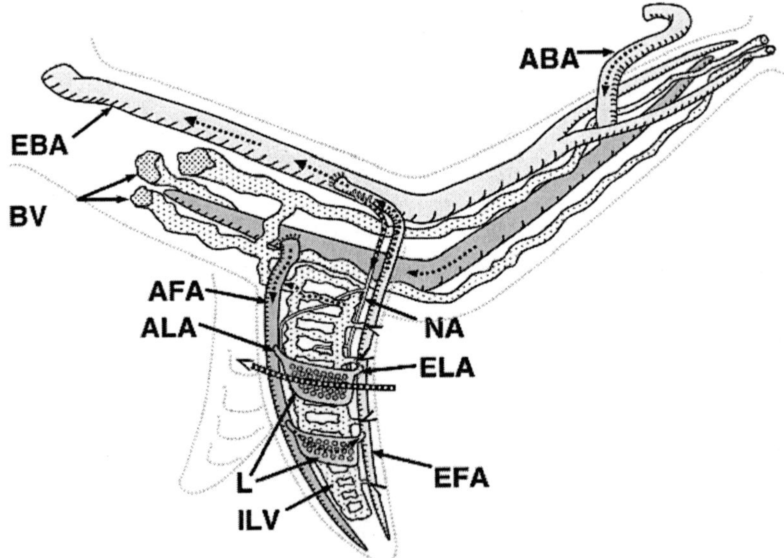

Figure: *Generalised schematic of blood flow through the major vessels of a gill arch and filament.* Arterio-arterial pathway*: blood travels (thin dotted arrows) from the afferent branchial artery (ABA) to an afferent filamental artery (AFA), which runs along the length of a filament. The AFA distributes blood to lamellae (L) via afferent lamellar arterioles (ALAs), and the lamellar blood is received by an efferent filamental artery (EFA) via efferent lamellar arterioles (ELAs). Blood flow through the lamellae is countercurrent to water flow across the lamellae (large white-on-black dotted arrow). The EFA returns blood from the filament to the efferent branchial artery (EBA), which distributes blood to the dorsal aorta for systemic circulation.* Arteriovenous pathway*: blood in*

the EFA can be distributed to the arteriovenous circulation, i.e., interlamellar vessels (ILVs), via postlamellar arteriovenous anastomoses (arrowheads) or by nutrient arteries (NA), which arise from the EFA or EBA. The ILVs are presumably drained by branchial veins (BV), which return blood to the heart.

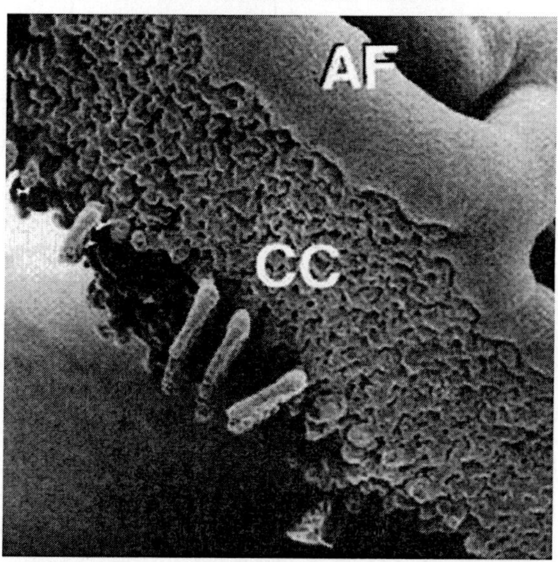

***Figure:** Scanning electron micrograph (x45) of a vascular cast from the gills of an elasmobranch (Western shovelnose stingaree,* Trygonoptera mucosa*), showing the afferent filamental artery (AF) and corpus cavernosum (CC).*

In teleosts, a CC is not present, but in some species the AFAs contain blebs or dilations, e.g., the channel catfish, *Ictalurus punctatus*; European perch, *Perca fluviatilis*; rainbow trout; and skipjack tuna, *Katsuwonus pelamis*. It has been hypothesized that these blebs are vestigial structures derived from the CC of elasmobranchs and agnathans or are functional structures that play a role in dampening pulsatile blood flow.

Regularly spaced along the length of the CC (elasmobranchs and lampreys) or AFA (teleosts) are afferent lamellar arterioles (ALAs) that feed one or more lamella(e). Blood flow through the lamellae is consistent with "sheet flow" theory and occurs through narrow vascular spaces delineated by pillar cells, termed the lamellar sinusoids.

Pillar cells are composed of a central trunk that houses the main body of the cell and thin cytoplasmic extensions (flanges) that spread from the trunk and connect to adjacent pillar cells, thereby lining the lamellar blood space.

In rainbow trout and another teleost, the roach (*Rutilus rutilus*), the lamellar sinusoids lead to deformation, deceleration, and redirection of erythrocytes during their passage through lamellae. This phenomenon is hypothesized to enhance lamellar gas exchange.

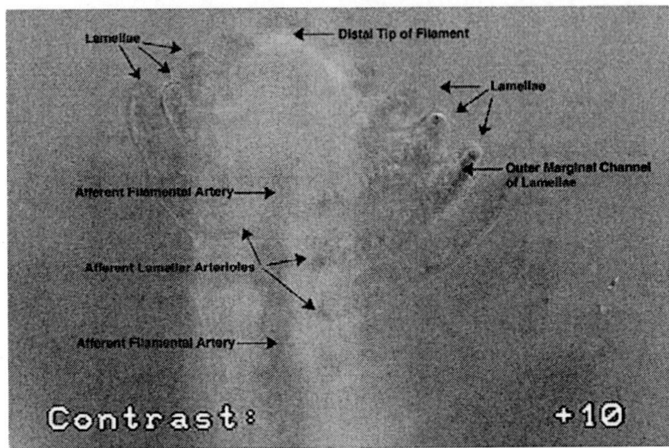

Figure: *Video of blood flow through the afferent filamental artery and a series of afferent lamellar arterioles leading into lamellae of a single filament in the gill of the American eel,* Anguilla rostrata.

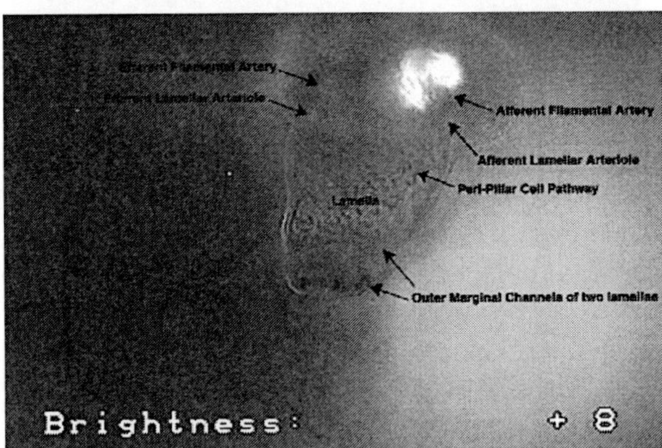

Figure: *Video of flow of blood through a single lamella in the gill of the American eel,* Anguilla rostrata.

Contractile microfilaments are found in pillar cells of several teleosts, lesser-spotted dogfish, and sea lampreys, and evidence for smooth muscle myosin has been detected in pillar cells from the Australian snapper (*Chrysophrys auratus*), a teleost. More recently, FHL5, a novel actin fiber-binding protein, has been localised in pillar cells in the Japanese eel, and its expression increased subsequent to acclimation to fresh water or volume expansion. These findings are suggestive of a contractile nature of pillar cells to possibly regulate blood flow through the lamellae. Recently, in the rainbow trout and Atlantic cod, in vivo videomicroscopy has provided direct evidence for regulation of blood flow through the lamellar sinusoids via pillar cell contraction.

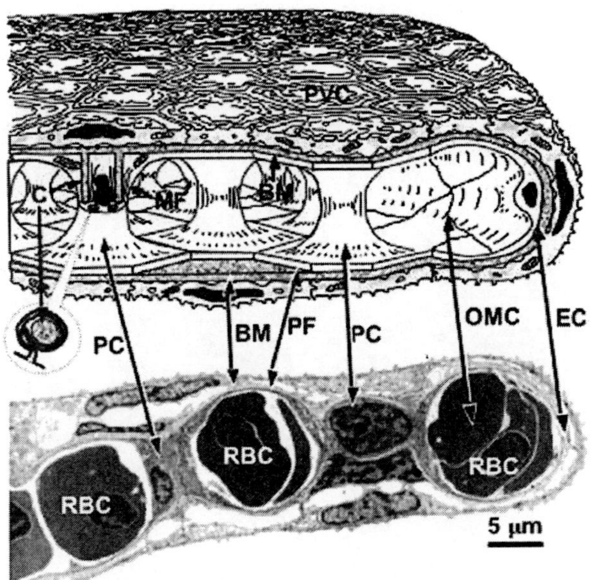

Figure: *Schematic and transmission electron micrograph of a cross section through an outer portion of a teleost (rainbow trout) lamella. Note the spool-shaped pillar cells (PC) with their cytoplasmic flanges (PF) that line the lamellar blood spaces, which are filled with red blood cells (RBC) in the micrograph. The pillar cells envelope extracellular bundles of collagen (C), which connect to the basement membrane (BM) that underlies the lamellar epithelium (e.g., PVC). Within the pillar cells are microfilaments (MF) that may be involved with pillar cell contraction. The outer marginal channel (OMC) is formed by pillar cells on its inner boundary and endothelial cells (ECs) on its outer boundary.*

In teleosts, elasmobranchs, and lampreys, pillar cells also envelop extracellular collagen bundles, which connect to collagen fibrils of the basement membranes that underlie the lamellar epithelial sheets. The collagen bundles may play a role in anchoring the opposing epithelial sheets to one another, and in maintaining structural integrity of a lamella during increases of blood pressure.

In addition to the lamellar sinusoids, blood flow across the lamellae may take a less convoluted route via the inner marginal channel (IMC) and outer marginal channel (OMC). The IMC is a discontinuous channel at the base of lamellae that is interrupted by pillar cells. This structure is embedded within the body of the filament body, which suggests it does not contribute to gas exchange. Moreover, evidence suggests that the IMC may allow blood to cross lamellae without exchanging gases with the respiratory medium, thus acting as a nonrespiratory shunt. The IMC may also be involved with the collection and distribution of blood within the lamellar sinusoids.

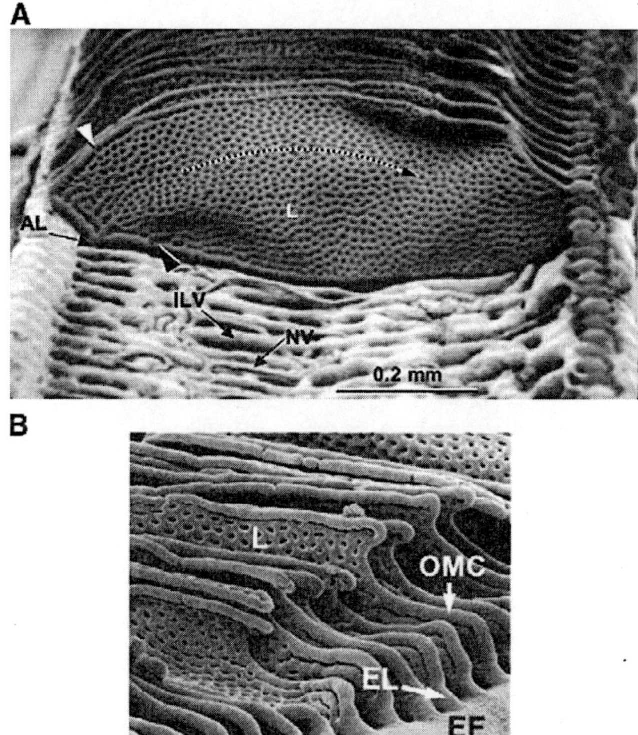

Figure: *Scanning electron micrographs of vascular casts from gill filaments of a teleost and an elasmobranch.* A*: a cast of a filament from a teleost (rainbow trout) with some lamellae (L) removed to show the underlying arteriovenous vasculature. Note the afferent lamellar arteriole (AL), inner marginal channel (black arrowhead), outer marginal channel (white arrowhead), an interlamellar vessel (ILV), and a nutrient vessel (NV). Dotted arrow indicates direction of lamellar blood flow. Holes in the lamellae correspond to the position of pillar cells.* B *(x125): a cast of a filament from an elasmobranch (sparsely spotted stingaree,* Urolophus paucimaculatus*) to show the efferent portion of lamellar blood flow. Note the outer marginal channel (OMC), efferent lamellar arteriole (EL), and efferent filamental artery (EF).*

In contrast to the IMC, the OMC is a low-resistance, continuous blood space at the periphery of lamellae, lined by endothelial and pillar cells on its outer and inner boundaries, respectively. Compared with the lamellar sinusoids, the OMC has a larger diametre, lower resistance, more erythrocytes, and greater surface area in contact with the ambient water; thus the OMC is hypothesized to be a preferential route of blood flow for shunting large volumes of blood around the lamellar sinusoids to maintain dorsal aortic blood pressure, and for possibly enhancing gas exchange. The latter hypothesis is supported by the recent data that demonstrate that the lamellae of skipjack tuna, a teleost with a high demand for oxygen, have multiple OMCs per lamella and a direct delivery of blood to the OMCs from the ALAs. In rainbow trout and Atlantic cod, the peptide hormone endothelin can redistribute lamellar blood flow to and from the OMC, which may allow these teleosts to alter the functional surface area of their lamellae. This may affect gas exchange as well as the diffusive gains and/or losses of water and osmolytes between a fish's blood and its environment.

Blood is collected from the lamellae by efferent lamellar arterioles (ELAs), which are short vessels that drain into the efferent filamental artery (EFA). The EFA travels along the entire length of a filament's efferent side, and blood flow through this vessel is counter to that of the AFA (i.e., toward the gill arch). The EFA also contains vessels that connect to the arteriovenous vasculature. Near the filament's origin on the arch, the EFA contains a distinctive muscular sphincter before it joins the EBA. This sphincter is highly innervated and may play a role in regulating lamellar blood flow and branchial resistance. The EBA collects oxygenated blood from the filaments and distributes the blood to the dorsal aorta for systemic circulation.

In the Atlantic and Pacific hagfishes, each gill pouch is supplied with blood by an ABA, which connects to an afferent circular artery (ACA) that encircles the excurrent duct of a pouch. Branching off the ACA are afferent radial arteries (ARAs) that feed each gill fold (i.e., filament) in a pouch. An ARA connects to cavernous tissue that permeates the afferent portion of the filament and is comparable to the CC of elasmobranchs and lampreys. At the efferent edge of the cavernous tissue, ALAs arise that provide blood to the highly folded gill lamellae. Interestingly, the complex folding pattern of the hagfish lamellae results in numerous OMCs.

Similar to other fishes, pillar cells within the lamellar folds of hagfish establish the lamellar sinusoids, and, like other fishes, the pillar cells contain microfilaments. Hagfish pillar cells can exist singly and envelope extracellular collagen bundles in a similar manner to other fishes, or pillar cells can occur in clusters of two or more cells that encircle extracellular collagen bundles. Intercellular junctions connect adjacent pillar cells in the Atlantic hagfish, which may suggest a degree of functional synchronisation between the cells, possibly to influence lamellar perfusion. These intercellular junctions are also found in the pillar cells of lampreys but have not been found in any other fishes.

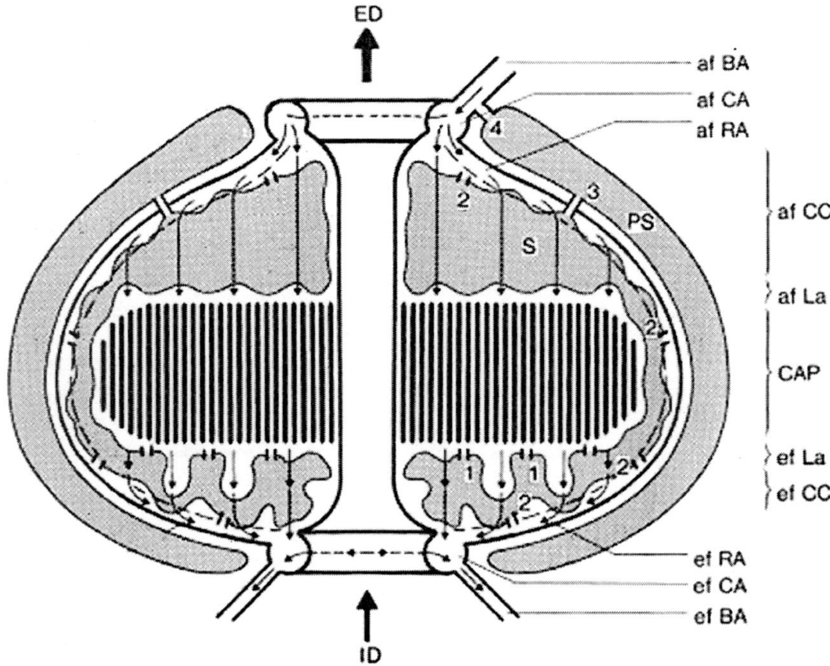

Figure: *Schematic of arterio-arterial and arteriovenous vasculature in the gill pouch and filament of hagfishes. The arterio-arterial vasculature includes afferent (af) and efferent (ef) branchial arteries (BA), circular arteries (CA), radial arteries (RA), cavernous tissue (CC), and lamellar arterioles (La). Also noted are the lamellar sinusoids or capillaries (CAP). The arteriovenous vasculature includes the peribranchial sinus (PS) and a sinusoid system (S). The numbers* 1, 2, *and* 4 *indicate sites of arteriovenous anastomoses. Number* 3 *marks an anastomosis between the venous sinusoid system and the venous peribranchial sinus. Thin arrows indicate direction of blood flow, and thick arrows indicate direction of water flow.*

Blood in the lamellar blood space of hagfishes is drained by ELAs that empty into an efferent cavernous tissue. This cavernous tissue is less complex than its afferent counterpart and feeds an efferent radial artery (ERA). The ERA then joins with an efferent circular artery (ECA) that encircles the pouch's incurrent duct. Blood from the ECA is returned to systemic circulation by two EBAs.

Arteriovenous Vasculature

The arteriovenous vasculature is often referred to as the nonrespiratory pathway. Although the exact function of this system is not known, it is likely involved with providing nutrients to the filament epithelium and the underlying supportive tissues of the gill filaments and may also provide a means for filamental blood to enter the venous circulation without traversing lamellae. The arteriovenous pathway may be considered a component of the systemic circulation because it is primarily supplied with postlamellar blood. Due to general similarities in this circulatory system among elasmobranchs and teleosts, these groups will be described together; the arteriovenous pathway of lampreys and hagfishes are each unique and thus are described separately. The core of the arteriovenous network in elasmobranchs and teleosts is composed of a highly ordered series of saclike vessels arranged like a ladder. The "rungs" of the ladder are termed the interlamellar vessels (ILVs), which lie underneath the interlamellar epithelium and run parallel to the lamellae. The "legs" of the ladder are termed the collateral vessels or sinuses, which flank and connect to the afferent and efferent boundaries of the ILVs, and run parallel to the length of the filament. The ILVs and their collateral vessels connect to other venous and sinus systems within the filament that return blood to the base of the filament and eventually to the heart.

In the literature, the ILVs are often collectively referred to as the central venous sinus (CVS), because descriptions of the ILVs morphologies vary among fishes and are conflicting. For example, vascular casts by some researchers in teleost and elasmobranchs found that the ILVs primarily appeared to be a continuous, singular, and amorphous sinus below the interlamellar epithelium that ran along the length of the filament, with little secondary structure, e.g., the rainbow trout; European perch; European eel, *Anguilla anguilla*; smooth toadfish, *Tetractenos glaber*; spiny dogfish; and Endeavour dogfish. In contrast, vascular casts by other researchers in some of the same and other teleost and elasmobranch species determined that the ILVs were primarily arranged like rungs on a ladder, e.g., the channel catfish; spiny dogfish; skipjack tuna; walking catfish, *Clarias batrachus*; Asian catfish, *Heteropneustes fossilis*; climbing perch, *Anabas testudineus*; lesser-spotted dogfish; spiny dogfish; little skate; sparsely spotted stingaree; and Western shovelnose stingaree.

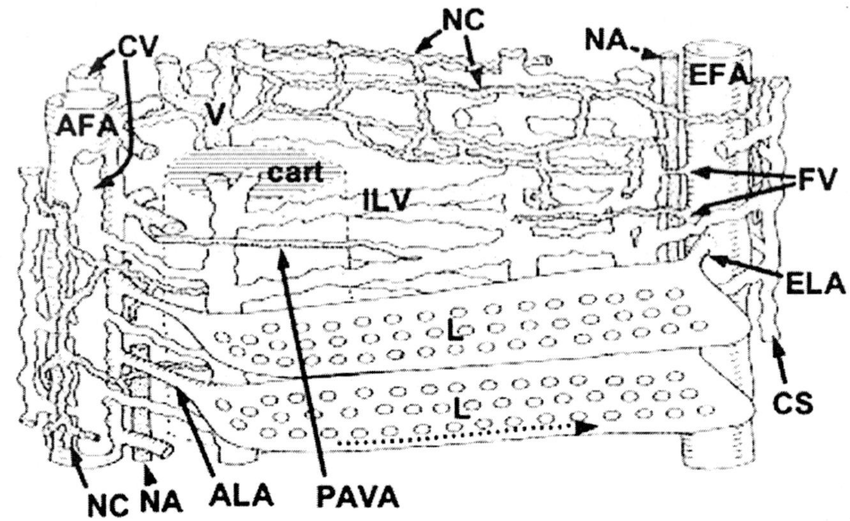

Figure: *Schematic of arterio-arterial and arteriovenous vasculature in the gill filament of teleosts. Nonstippled vessels represent the arterio-arterial vasculature, i.e., afferent filamental artery (AFA), afferent lamellar arterioles (ALA), lamella (L), efferent lamellar arterioles (ELA), and efferent filamental artery (EFA). Dotted arrow indicates direction of lamellar blood flow. Stippled vessels represent the arteriovenous vasculature, i.e., interlamellar vessels (ILV), collateral vessels (CV) and sinuses (CS), prelamellar arteriovenous anastomoses (PAVA), postlamellar arteriovenous anastomoses or feeder vessels (FV), nutrient arteries (NA), nutrient capillaries (NC), and filament veins (V). Also note filamental cartilage (cart) that runs along the length of the filament on the afferent edge.*

Because the morphology of these vessels in the rainbow trout is sensitive to the perfusion pressure of the vascular casting agent, the latter description is likely the true structure of the ILVs. Therefore, the continuous, amorphous morphology of the ILVs described by some researchers was probably an artifact of superphysiological perfusion pressures, which greatly distend the vessels and create the appearance of a singular sinus. Regardless, the findings of the above-mentioned studies suggest that the ILVs are pliable vessels with an ability to change their structure, and possibly their function, in response to changes in blood pressure.

Blood is supplied to the arteriovenous vasculature by at least three sources: *1*) prelamellar arteriovenous anastomoses (AVAs), *2*) postlamellar AVAs, and *3*) filamental nutrient vessels. Prelamellar AVAs connect the AFA, ALA, or CC to the ILVs. These AVAs are not universal among teleosts and elasmobranchs but have been detected in several species, e.g., channel catfish, smooth toadfish, European eel, Endeavour dogfish, and spiny dogfish. In contrast, postlamellar AVAs (also referred to as feeder vessels) that connect the EFA and/ or ELA to the ILVs occur much more frequently than prelamellar AVAs and have been detected in the vast majority of teleost and elasmobranch species examined to date.

It is important to note that blood in the ILVs of teleosts is often characterised by a lower hematocrit than blood found in the arterio-arterial circulation [e.g., the rainbow trout, channel catfish, and black bullhead. These findings suggest that the prelamellar and postlamellar AVAs may be sites of plasma skimming, which emphasizes the nonrespiratory function of the ILVs.

The demonstration of both prelamellar and postlamellar AVAs in some species was consistent with early experimental data that suggested the ILVs may function as a putative lamellar bypass for blood, i.e., a pathway for blood to pass from the AFA (or CC) to the EFA without traversing the lamellar sinusoids. However, this hypothesis was largely refuted, and it is more likely that prelamellar AVAs act as a lamellar shunt, i.e., a pathway for blood to pass from the AFA (or CC) to the ILVs for venous return to the heart without traversing the lamellar sinusoids.

Another source of blood to the ILVs is through nutrient vessels, which compose another meshwork of vasculature within the core of the gill filament termed the nutrient system (NS).

It is unclear, however, whether the NS represents a distinct subsystem of the arteriovenous vasculature or is simply another source of blood to the ILVs, like AVAs. Given the infrequent observations of nutrient vessels connecting to the ILVs, it is more likely that the NS is in fact a discrete component of the arteriovenous vasculature that supplies nutrients to the underlying tissues of the filaments and arches and is not just a supplier of blood to the ILVs.

Arterioles and arteries that bud off the EFA and EBA primarily supply blood to the NS. In teleosts, it is common for short, but often tortuous, small-diametre arterioles to branch off the EFA and EBA, and then anastomose into larger diametre nutrient arteries, e.g., rainbow trout, European perch, climbing perch, walking catfish, and skipjack tuna.

In elasmobranchs and some teleosts, large-diametre nutrient arteries arise directly from the EFA, e.g., spiny dogfish, little skate, smooth toadfish, and channel catfish. Regardless of how nutrient arteries originate, they travel along the filament's length alongside the EFA or may bifurcate and traverse towards the afferent side of the filament where they branch into nutrient arteries that run medially along the filament's long axis, or into nutrient arteries that run adjacent to the AFA.

Additionally, smaller vessels may branch off the nutrient arteries towards the centre of the filament and anastomose with one another to form a weblike arrangement of nutrient arterioles and capillaries that surround, and may anastomose with, the ILVs.

Given the similar location and trajectories of the ILVs and some vessels of the NS, it is often difficult to discern one system from the other with vascular casts, especially in cases when the ILVs are distended and envelop parts of the NS.

However, there are unique morphological features of the vessels that identify these two systems. For example, the vessels of the NS tend to have typical artery/arteriole morphology with a robust endothelium and many erythrocytes, whereas vessels of the ILVs have a more pliable venouslike morphology with a meager endothelium and few erythrocytes. Regardless, both the ILVs and NS are drained by venous systems within the filament that return blood to the base of the filament and eventually to the heart.

In lampreys, the arteriovenous vasculature has not been extensively studied. However, a structure in a similar anatomical location to teleost and elasmobranch ILVs, termed the "axial plate lacunae," has been described in the Arctic lamprey. This structure is openly connected to the CC and EFA without AVAs and appears to function more like an IMC of a teleost lamella than an ILV. The major venous drainage structures in lamprey filaments are the peribranchial sinuses (PBS), which are located at the basal portion of hemibranchs, on both sides of an interbranchial septum.

Blood is supplied to the PBS by filament veins, which run parallel to and alongside the AFA. These veins receive blood from the AFA via AVAs that are analogous to prelamellar AVAs of elasmobranchs and teleosts; filament veins then feed the PBS via anastomosing channels along the vein's length. The PBS is also supplied with postlamellar blood via AVAs in the EFA that are analogous to postlamellar AVAs of elasmobranchs and teleosts. From the PBS, blood eventually returns to the heart via other sinuses and veins. It is not known if nutrient vessels are found in the arteriovenous vasculature of lampreys.

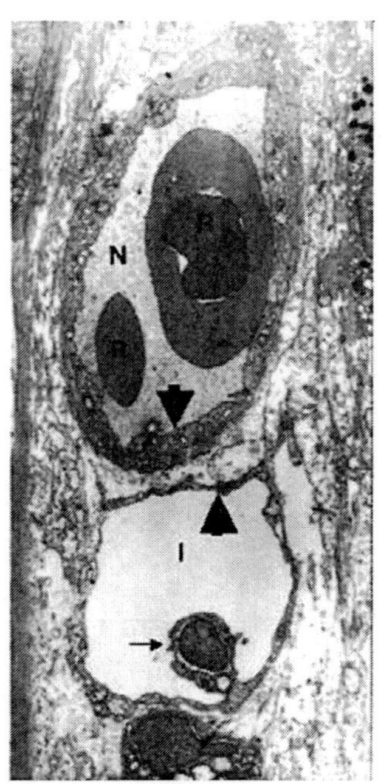

Figure: *Transmission electron micrograph (x4,400) of a section through a nutrient (N) and interlamellar (I) vessel. Note the thicker endothelium (large arrows) and occurrence of red blood cells (R) in the nutrient vessel, compared with the interlamellar vessel. Small arrow indicates a nucleus bulging from an endothelial cell.*

In hagfishes, the arteriovenous vasculature also has not been extensively studied, but evidence suggests it shares some similarities with that of the otherfish groups. Similar to lampreys, a PBS is present, which surrounds the gill pouch. In the Atlantic hagfish, a sinusoid system, which is analogous to the ILVs of elasmobranchs and teleosts, surrounds the afferent and efferent portions of the filaments, but not the lamellae. In the afferent and efferent cavernous tissues, AVAs supply prelamellar and postlamellar blood, respectively, to the sinusoid system. Other AVAs also have been noted between the sinusoids and the radial vessels, but it is not clear if nutrient vessels are present.

Neural

Branchial nerves in fishes are derived from the VIIth (facial), IXth (glossopharyngeal), and Xth (vagus) cranial nerves, each divided into posttrematic, motor rami, and pre- and posttrematic, sensory rami. In a given gill arch, posttrematic rami from one nerve may mix with pretrematic rami from an adjacent nerve. Gill arches other than the first are generally innervated by the vagus nerve. There are substantial data supporting sensory and motor function for the teleost branchial autonomic innervation; however, similar functional data for elasmobranchs and agnatha have not been published, despite the fact that at least adrenergic and cholinergic receptors have been functionally described in the branchial vasculature of both elasmobranchs and hagfish.

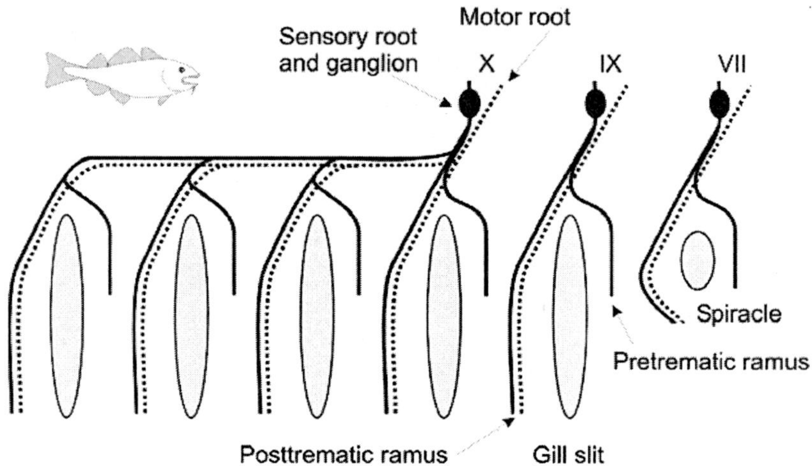

Figure: *Outline of the components of the branchial nerves of fish, showing facial (VII), glossopharyngeal (IX), and vagal (X) rami in relation to the gill slits. The orientation is cephalic to the right, as indicated by the outline of the cod.*

Histochemical studies have described catecholamine-containing fibres associated with AFAs and ALAs in a variety of teleost species, and similar fibres associated with ELAs and EFAs in some species, but the lamellae do not appear to be innervated. On the other hand, catecholamine-containing fibres have been described around the ILV in a number of teleost species. Cholinergic-type fibres have been described in the gills but are only associated with the efferent vasculature, specifically with a prominent sphincter in the basal third of the efferent filamental artery, which also appears to contain serotonergic fibres. Serotonergic neuroepithelial cells (NECs) have also been described on the gill filament in a variety of species, as have fibres and NECs that contain immunoreactive nitric oxide synthase. Specific roles for the neurosecretory products of these fibres and cells are discussed in sections IV*D* and VIII*B*.

Gas Exchange and Gas Sensing

Because the branchial epithelium is essentially a monolayer of mixed cells (mostly squamous PVCs on the lamellar surface), and the medium irrigates the external surface countercurrent to the internal perfusion, the fish gills appear to be ideal for gas exchange. The ratio of irrigation (ca. 5,000–20,000 ml $\cdot$ kg^{-1} $\cdot$ h^{-1}) to perfusion (Vg / Q) is generally of the order of 1–8 but may approach 50 in very active, pelagic fishes such as tuna (by substantial increases in $_g$), and oxygen extraction efficiencies vary from 50 to 90%, higher than those described for mammals.

The metabolic cost of gill gas exchange is substantial, however, because of the low solubility of oxygen in aqueous solutions (~7 ml/l; 3% that in air) and the high viscosity (800 times air) and density (60 times air) of the medium. Thus the cost of routine ventilation may be 10% of the oxygen uptake (O_2), and it may approach 70% of MO_2 during exercise, when ventilation volumes may triple. Pelagic fish species reduce the cost of ventilation by utilising forward movement and open mouths to drive water across the gills.

Lamellar Gas Exchange

The actual site of gas exchange is assumed to be the lamellar epithelium, but no direct measurements of lamellar versus filamental gas permeabilities have been published. The lamellar surface is likely the primary site, because various studies have shown a correlation between lamellar surface area and respiratory needs (e.g., benthic vs. pelagic) or measured rates of oxygen uptake.

In fact, recruitment of distal lamellae is generally considered to be a means of increasing gas exchange, although this hypothesis is largely based on two studies that demonstrated that only 60–70% of lamellae in rainbow trout and lingcod (*Ophiodon elongatus*) were perfused at rest. In the trout, lamellar perfusion was measured by the distribution of vitally stained erythrocytes after infusion into trout via a cannula in the ventral aorta; in the lingcod, the distribution was measured by perfusion of isolated branchial arches with saline containing Prussian Blue dye.

In both cases, it is unclear if lamellar perfusion mimicked the in vivo condition. Our own videography of blood flow through the gills of American eels (*Anguilla rostrata*) or longhorn sculpin (*Myoxocephalus octodecimspinosus*) indicates that even distal lamellae are perfused in anaesthetised individuals, so lamellar recruitment may not be a general phenomenon. The respiratory (lamellar) surface area is generally 0.1–0.4 m^2/kg but may be as high as 1.3 m^2/kg in tuna (equivalent to human) or as low as <0.1 m^2/kg in species that utilise the swim bladder (or other accessory air-breathing organs) for oxygen uptake. The respiratory medium (water) to blood diffusion distance (i.e., the thickness of lamellar epithelium) also varies with life-style, ranging from 10 μm in aerial breathers to <1 μm in tuna. It is not clear if this dimension can be dynamically regulated, but thinning of the epithelium might possibly occur if an increase in cardiac output distends lamellae. In addition, alteration in the distribution of gill cells (MRC vs. PVC) under hypoxic or hypercapnic conditions may alter the diffusion distance and thereby impact gas exchange.

Perfusion Versus Diffusion Limitations

The uptake of oxygen across the fish gills appears to be perfusion limited, but all data published to date are from a single species, the rainbow trout, so more data are needed to support the general statement. Nevertheless, the fact that rainbow trout arterial oxygen content (Pa_{O2}) did not vary appreciably with changes in perfusion produced by alteration of blood volume argues for perfusion limitation, as does the lack of any alteration in Pa_{O2} subsequent to the ligation of two gill arches. In contrast, excretion of CO_2 by the rainbow trout gill epithelium appears to be diffusion limited, at least in teleosts. Thus alterations in blood volume in the trout were correlated with alterations in blood CO_2, and in rainbow trout a 30% reduction in gill surface area was accompanied by a significant elevation of Pa_{CO2}. For a more complete discussion of the data supporting perfusion limitation for O_2 uptake and diffusion limitation for CO_2 excretion across the fish gills.

The diffusion limitation to CO_2 excretion in teleosts is apparently secondary to the need for erythrocyte Cl^-/HCO_3^- exchange (via band 3) at the gills, the only means of CO_2 production at the gas exchanger because of the lack of endothelial carbonic anhydrase (CA), as is found in the mammalian lung alveoli. The recent findings that injection of CA offset the increased Pa_{CO2} seen in the rainbow trout after an increase in cardiac output or decrease in gill surface area support this conclusion. It is unclear if the constraints of CO_2 excretion are the same in most fishes, but it is apparent that the dynamics of CO_2 excretion across the gills in at least three other fish taxa may be different. For instance, elasmobranchs do have vascular CA, hemoglobin-less icefishes lack erythrocytes and therefore band 3, and lampreys are unique among vertebrates in the lack of a functional anion exchanger in the erythrocyte. One might expect CO_2 excretion in these three groups to be perfusion limited, but this hypothesis has not been tested to date. Despite these theoretical constraints on CO_2 excretion, the fish gills appear to be hyperventilated with regard to CO_2, because the blood Pv_{CO2}, and bicarbonate concentration (~4 mM) is far below that found in mammals, producing a plasma pH (~7.8) that is substantially higher than that measured in mammalian plasma. The impact of these values on acid-base regulation is discussed in section VI*B*.

Chemoreceptors

O_2 Sensors

Because the branchial arch arteries of fishes are the evolutionary precursors of the carotid and pumonary arteries in mammals (402), one might suspect that the gills would be the site of afferent sensory neurons whose efferent motor output may control cardiovascular and respiratory responses to environmental and/or systemic changes in dissolved gas concentrations.

This is, in fact, the case. As is found in other vertebrates, changes in environmental or blood gas tensions elicit a suite of ventilatory and cardiovascular responses in fishes. The primacy of O_2as the proximate signal is well established, but recent studies have demonstrated that CO_2/pH also may be a relevant signal. Environmental hypoxia elicits hyperventilation in a variety of species, and the response is generally larger changes in stroke volume than rate, thereby reducing the energetic effects of irrigating the gills with a relatively dense and viscous medium; hyperoxia elicits the opposite response.

The responses occur within seconds of the exposure, suggesting external receptors, and this hypothesis is supported by the finding that externally applied cyanide elicits a hyperventilatory response. The receptors are distributed throughout the gill arches and are innervated by the 9th and/or 10th cranial nerves. There are also internally oriented, oxygen chemoreceptors in the gill vasculature that respond to either hypoxemia (in the absence of external hypoxia) or injection of cyanide. Such receptors would be advantageous during exercise or other changes in metabolic rate. It is not clear whether the external or internal chemoreceptors respond to Pa_{O2} or oxygen content, although it is more likely that tension is the proximate signal since vagus nerve recordings from isolated gill arches from both tuna and rainbow trout showed responses to reductions in perfusate Pa_{O2} even when total O_2 concentration was unchanged. Extrabranchial, buccal O_2-sensitive chemoreceptors have also been described, in particular in air-breathing fishes, but there is no convincing evidence to date for O_2 receptors in the central nervous system of fish.

The neuroendocrine physiology of gill oxygen chemoreceptors is relatively unstudied, although current evidence suggests that neuroepithelial cells (NEC) are involved, as they are in the lungs of terrestrial vertebrates. Dunel-Erb et al. described NEC on the gill filaments (near the efferent filamental artery, facing the afferent respiratory water flow) of a variety of teleosts and one elasmobranch, that degranulated under hypoxic conditions, and serotonergic NEC have been confirmed in other species.

More recently, it has been shown that zebrafish (*Danio rerio*) respond to hypoxia, and the filaments and lamellae of all four gill arches of the developing embryo contain both serotonergic and nonserotonergic NEC. Innervation patterns in this preparation suggest that the serotonergic NEC may be generating signals back to the central nervous system (to modulate respiratory and cardiovascular parametres) as well as releasing paracrines to affect local perfusion patterns within the filament and lamellae. The nonserotonergic NEC are immunopositive for the purinergic , not innervated, and often associated with the pillar cells of the lamellae, suggesting a paracrine role for these NEC.

CO_2 Sensors

The general finding that ventilation is matched to oxygen demands, even at the expense of blood pH (e.g., hyperventilation in hypoxic conditions is accompanied by respiratory alkalosis), suggests that O_2 demand overrides acid-base disturbances. This does not mean, however, that a CO_2/pH ventilatory drive does not exist. For instance, increasing the Pa_{CO2} of a hyperoxic medium abolishes the usual hypoventilatory response in rainbow trout or may even produce hyperventilation in the channel catfish.

Indeed, the physiology of putative CO_2/pH-sensitive receptors in the gill epithelium has generated much recent interest. Fishes from all major taxonomic groups respond to environmental hypercapnia by a significant increase in gill irrigation and bradycardia. Until recently, it was assumed that the responses were indirect, the result of hypoxemia produced by the effect of blood hypercapnia on hemoglobin oxygen affinity (Bohr effect) and carrying capacity. More recent studies, however, have demonstrated a direct effect of CO_2/pH on cardiorespiratory parametres, under normoxemic conditions.

The CO_2/pH-sensitive chemoreceptors appear to be primarily on the first gill arch in some species but may be more generally distributed in other species, because the responses are abolished only when all gill arches are denervated. The receptors appear to respond to external rather than internal changes in CO_2/pH, because injection of acetazolamide (which raises blood Pv_{CO2}) did not alter ventilatory frequency or amplitude in the rainbow trout, catfish, and spiny dogfish, nor did a bolus infusion of CO_2-enriched saline into the posterior caudal vein of the spiny dogfish. On the other hand, the fact that postexercise hyperventilation in the rainbow trout was reduced by injection of CA (which reduced the attending respiratory acidosis) suggests that internal receptors may be present.

In all tetrapod vertebrates, receptors in the central nervous system respond to changes in perfusate CO_2/pH, but it is unclear if such receptors are present in the central nervous system of water-breathing fishes. Forty years ago, an experiment on a single tench (*Tinca tinca*) found that injection of slightly acidic saline into the posterior medulla increased the amplitude of electrical discharges that were correlated with respiratory movements, and a subsequent study demonstrated an inverse correlation between the bicarbonate concentration of the saline perfusing the isolated brain of the lamprey (*Lampetra fluviatilus*) and the frequency of putative respiratory motor output. Recent studies of water-breathing fishes have not corroborated these findings.

For instance, hypercapnia-induced hyperventilation in the skate (*Raja ocellata*) was correlated with the pH of the arterial blood, not the cerebrospinal pH or intracellular pH of brain tissue, and similar findings have now been reported for other species. On the other hand, aerial gas exchange has arisen multiple times in the piscine lineage (e.g., gars, eels, tarpon, lungfishes), and it appears that central CO_2/pH chemoreceptors have evolved in at least some of these groups. For example, reduction of the pH of the solution perfusing the brain stem-spinal cord preparation of the gar (*Lepisosteus osseus*) was correlated with a significant increase in the putative air-breathing motor output, but not the motor output thought to be correlated with gill ventilation. Similar results have been described for the South American lungfish (*Lepidosiren paradoxa*), where lowering the pH of the perfusate in the IVth ventricle was associated with an increase in lung ventilation frequency. Thus the extant data do not allow us to stipulate exactly when central CO_2/pH receptors evolved in fishes, but most likely it was associated with either the independent evolution of aerial respiration in early fish lineages (represented by gar) or with the lineage that gave rise to the amphibia (represented by lungfishes). Clearly more data are needed on this interesting question.

A fall in intracellular pH is generally considered to be the proximate stimulus in the mammalian carotid chemoreceptor, and it is likely that this is also the intracellular stimulus in the fish chemoreceptor. Recent evidence suggests that the proximate, extracellular stimulus in the fish gill chemoreceptor is actually a change in extracellular CO_2 rather than pH. For instance, changes in medium pH did not elicit the usual cardiorespiratory responses of two tropical teleosts if the Pv_{CO2} was unaltered. In the Atlantic salmon and spiny dogfish, CO_2-enriched water produced the expected bradycardia and hyperventilation, but acidified, CO_2-free water did not elicit cardiorespiratory changes.

In summary, respiratory function in fishes (using the gills) is similar to that in tetrapod vertebrates (using lungs), but medium-constrained differences do exist. There is a need for more studies in a variety of fish groups before some of the generalities presented above can be verified.

Osmoregulation and Ion Balance

The structure and function of the branchial epithelium as a site of gas exchange dictates that it is also the site of osmosis and ionic diffusion if gradients exist. The marine, invertebrate ancestors of the vertebrates (i.e., sister taxa to modern echinoderms and tunicates) were and are isosmotic (but not always isoionic) to seawater. In contrast to an early hypothesis of freshwater origin, it is now generally accepted that the first vertebrates evolved in seawater, entered brackish and fresh water (one lineage giving rise to the Amphibia), and then reentered the marine environment in some cases. The suggestion that the presence of a renal glomerulus in nearly all vertebrates is the result of origin in fresh water is invalidated by the presence of functional glomeruli in the totally marine hagfish, modern members of the earliest fish lineage, which do not have a history of freshwater ancestry. The scenario of marine origin followed by entry into fresh water also is supported by the extant fossil record and is the most parsimonious explanation for the isotonicity of hagfish to seawater and the general finding that the NaCl concentration of the blood and extracellular fluids of all the other fish taxa (indeed, all other vertebrates) is ~30% that found in seawater, but distinctly above the NaCl content of fresh water. The maintenance of blood and intracellular tonicity in the face of these osmotic and ionic gradients across the fish branchial epithelium takes energy, but the most recent calculations suggest that it is only of the order of 2–4% of the resting oxygen consumption, far below the cost of ventilation.

Apical Na^+ Uptake

The mechanisms of ionic uptake were first suggested by August Krogh, who found that Na^+ and Cl^- were extracted from fresh water independently by the goldfish (*Carassius auratus*) and suggested that blood ions such as NH_4^+, H^+, and HCO_3^- could function as counterions to maintain electroneutrality across the epithelium. Subsequent studies appeared to support this hypothesis, but uptake of Na^+ via passive exchange with either NH_4^+ or H^+ has been questioned on thermodynamic grounds.

For instance, gill epithelium intracellular Na^+ concentrations have been reported in the range of 50–90 mM in two species of trout and ~10 mM in tilapia, far above the common freshwater concentration of <1 mM Na^+. In addition, it has been calculated that a pH gradient of 0.3 units would be necessary to drive H^+ outward (in exchange for external Na^+), but rainbow trout are able to excrete protons into water with a pH of 6 despite the fact that the gill epithelial cell pH in this species is ~7.4, and blood pH is even higher. In addition, many of the early studies that appeared to support the model of Na^+/H^+ or Na^+/NH_4^+ exchange by demonstrating sensitivity of Na^+ uptake to amiloride used concentrations (0.1–1 mM) that could not distinguish among inhibition of ionic exchange, blockade of an apical Na^+channel, or even inhibition of basolateral Na^+-K^+(NH_4^+)-ATPase. Moreover, the Na^+/H^+ exchanger (NHE)-specific inhibitor dimethylamiloride did not inhibit Na^+ uptake by the isolated European flounder (*Platichthys flesus*) gill filament.

Recent data suggest that the uptake of Na^+ by the fish gill epithelium may be electrically coupled to proton efflux, rather than directly coupled by a single exchanger (i.e., NHE) to proton excretion. For instance, it was found that the HCO_3^- concentration in the external water (at a neutral pH) actually fell as the medium flowed past the rainbow trout gills, contrary to what one might expect if excreted CO_2 was hydrated in the unstirred layer on the water side of the gill epithelium. Moreover, the fall in HCO_3^- concentration was not altered by the disulfonic stilbene SITS, so it was probably not due to reversed Cl^-/HCO_3^- exchange. These data suggest that secreted protons dehydrate the HCO_3^- in the unstirred layer.

Addition of vanadate to the external medium at a concentration (0.1 mM) that does not inhibit basolateral Na^+-K^+-ATPase in the frog skin inhibited 50% of the putative proton secretion by the rainbow trout gills, suggesting that the excretion was via an apical, P-type proton pump. Other studies described a vesicular fraction isolated from rainbow trout gills that transported protons and expressed a Mg^{2+}-dependent ATPase that was sensitive to *N*-ethylmaleimide, a known inhibitor of both V-type and P-type H^+-ATPases, but with a 10- to 100-fold selectivity for the former. Moreover, one study was in the presence of EGTA, azide, and ouabain, so activity of Ca^{2+}-activated ATPase, mitochondrial H^+-ATPase, and Na^+-K^+-ATPase was assumed to be minimal. In addition, the NEM-sensitive ATPase activity was also inhibited by other H^+-ATPase inhibitors, such as dicyclohexylcarbodiimide (DCCD), diethylstilbestrol (DES), *p*-chloromercuribenzenesulfonate (PCMBS), and bafilomycin. The inhibitor profile did not allow a definitive differentiation between P- and V-type proton pumps, but subsequent physiological and immunological studies have made a V-type proton pump (V-H^+-ATPase) the most likely. For example, the V-H^+-ATPase specific inhibitor bafilomycin (1–10 μM in the freshwater medium) inhibited 60–90% of Na^+ uptake by tilapia, carp (*Cyprinus carpio*), and zebrafish. These results corroborate molecular studies that showed that heterologous antibodies raised against various subunits of V-H^+-ATPase localised expression to the branchial epithelium of rainbow trout, coho salmon, tilapia, zebrafish, and two elasmobranchs: the spiny dogfish shark and Atlantic stingray.

Homologous antibodies raised against the zebrafishisoforms demonstrated expression in gill tissue from that species. In addition, the mRNA for V-H^+-ATPase has been localised in the branchial epithelium of the rainbow trout by Northern blot, using a probe complementary to the cloned B-subunit of the rainbow trout gene as well as by in situ hybridisation using probes complementary to the bovine renal proton pump. The amino acid sequence for this rainbow trout gill V-H^+-ATPase subunit shares >95% identity with the vatB1 isoform of the eel swim bladder gas gland V-H^+-ATPase and zebrafish gill vatB1, as well as the vatB1 isoform from humans. Similar amino acid sequence identities have been shown for the vatB2 isoform from the European eel gas gland and zebrafishgills compared with the human sequence, but the gene for the vatB2 isoform has not been cloned from the rainbow trout gill. A partial sequence from the - subunit from the gill of the Osorezan dace (*Tribolodon hakonensis*) has been reported, as well as a full-length clone of the á-subunit of V-H^+-ATPase from the killifish gill.

The presence of V-H^+-ATPase in the fish branchial epithelium supports the hypothesis that Na^+ uptake by this tissue may be coupled to electrogenic proton extrusion, as has been described for tight epithelia such as frog skin and distal portions of the mammalian nephron. The multistranded junctional complexes between MRC and PVC cell in the branchial epithelium of freshwater fishes suggest an electrically tight epithelium, and the extremely high resistance of the cultured, freshwater rainbow trout gill epithelium (3.5 kÙ · cm^2) supports this proposition. If the "tight epithelium" model of electrical coupling between proton extrusion and Na^+ uptake is correct, then V-H^+-ATPase and a Na^+ channel should be colocalised to the apical membrane, and Na^+-K^+-ATPase should be localised to the basolateral membrane of the same cell. The initial immunolocalisation of V-H^+-ATPase suggested that both MRC and PVC in the rainbow trout gill expressed the proton pump, and the labelling in both cells now has been shown to be apical. In the coho salmon, V-H^+-ATPase was localised to MRC. Another study on the rainbow trout, however, has localised the proton pump in the apical region of only PVCs, and this PVC-only localisation has now been extended to tilapia.

This localisation of the proton pump to the PVC was also supported by PVC versus MRC distribution changes under acidotic conditions, high-magnification transmission electron microscopy, and X-ray microanalysis studies. Whether these differences in the cellular localisation of V-H^+-ATPase are antibody specific or due to differences in rainbow trout populations, holding conditions, or species, remains to be seen, but current data do support an apical position for the proton pump in one or more types of epithelial cells in the freshwater fish gill. An amiloride-sensitive Na^+ channel has not been cloned from fish, but benzamil (an amiloride analogue with a high affinity for ENaC) inhibited Na^+ uptake by freshwater European flounder gills. In addition, a recent study demonstrated that an ENaC polyclonal antibody (raised against a synthetic peptide corresponding to residues 411–420 of the -subunit of the human epithelial Na^+channel) colocalised with the V-H^+-ATPase, in only PVC in tilapia and in both PVC and MRC in the rainbow trout.1 In all of these studies, MRC cells were identified either by their morphology (and/or position in the epithelium) or by immunological staining with heterologous antibodies raised against Na^+-K^+-ATPase, which is assumed to have high expression in the MRC. Recent cell separation techniques may offer an avenue for better delineation of specific cellular function. It has been demonstrated that PVC and MRC from the rainbow trout gill can be separated by Percoll density gradient, combined with peanut lectin agglutinin (PNA) binding and magnetic bead separation.

With the use of this technique, MRC can be separated into two subpopulations, one of which (PNA negative; termed á-MRC) displays bafilomycin-sensitive, acid-activated ^{22}Na uptake, which also is sensitive to the amiloride analogue phenamil, an ENaC inhibitor. Importantly, in this study, the PVC (defined by position in the Percoll gradient) displayed very low rates of Na^+ uptake that were not sensitive to bafilomycin.

This suggests that the acid-secreting cells in the rainbow trout gill are a subpopulation of MRCs, not PVCs. Thus data supporting the tight epithelium model for Na^+ uptake (and H^+ extrusion) exist for the branchial epithelium of freshwater teleosts, but the data are restricted to two species (rainbow trout and tilapia) and must be extended to a wider array of species. This model for Na^+ uptake across the apical membrane of the freshwater fish gill epithelium may apply to Na^+ uptake from very dilute solutions in some teleosts. But passive, chemically coupled Na^+/H^+ exchange could mediate Na^+ uptake if electrochemical gradients for either Na^+ or H^+ are favourable (e.g., in solutions with a $[Na^+]$ higher than a few mM and/or pH >7.6, which may be found in some fresh water).

Under these circumstances, one might propose apical Na^+/H^+ and Cl^-/HCO_3^- exchangers coupled to basolateral Na^+-K^+-ATPase. These are the apical transporters responsible for Na^+ absorption in mammalian intestinal and renal proximal tubule cells and therefore would be expected to be the isoforms if apical NHEs function in freshwater fishes. Antibodies raised against human proteins have been used to localise NHE3 immunoreactivity in the branchial epithelium of freshwater rainbow trout and NHE2 immunoreactivity in the branchial epithelium of freshwater tilapia.

Neither study specifically localised Na^+-K^+-ATPase to differentiate between MRC and PVC, but morphology and position of the stained cells suggested that the NHEs were only in PVC in the rainbow trout and in both PVC and MRC in tilapia. The most convincing evidence that an apical NHE is expressed in gills in freshwaterfishes comes from a recent study that cloned a full-length transcript homologous to NHE3 from the gills of the Osorezan dace and used a homologous antibody to localise expression to the apical surface of MRC.

Data from another species, the killifish, suggest a third alternative for the distribution of the pumps and channels necessary for at least Na^+ uptake by the fish branchial epithelium. V-H^+-ATPase has been localised to the basolateral membrane of cells that also express Na^+-K^+-ATPase, using homologous antibodies (raised against a peptide fragment coded by cloned, killifish V-H^+-ATPase).

This suggests that basolateral cation extrusion by both pumps may produce an electrochemical gradient across the apical membrane that is sufficient to drive Na^+ into the cell from the external fresh water.

It is not clear if environmental or species differences can account for these disparate models for Na^+ uptake across the freshwater fish gill epithelium, but it is clear that molecular techniques, applied to a variety of species and conditions, are needed to provide a better delineation of the mechanisms of Na^+ uptake.

Apical Cl^- Uptake

Chloride uptake by the freshwater fish gill generally is considered to be via Cl^-/HCO_3^- exchange, but only a few studies have examined this putative pathway directly.

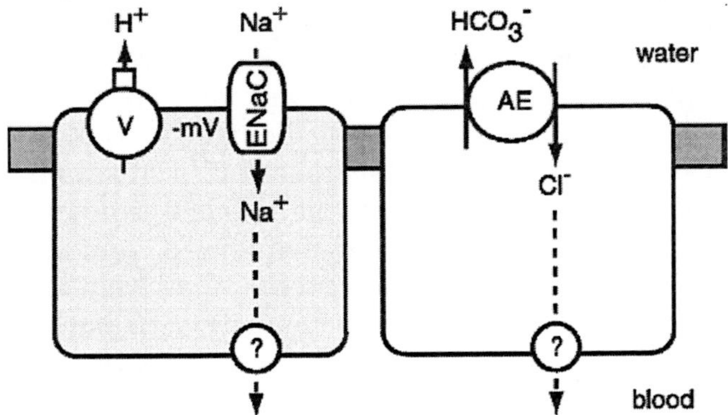

FW Salmonids and Tilapia

Figure: *Working model of NaCl uptake mechanisms proposed for freshwater Atlantic stingrays. One type of MRC (A MRC) expresses Na^+-K^+-ATPase on its basolateral membrane and is hypothesized to draw in Na^+ across the apical surface in exchange for cytoplasmic H^+. The other MRC (B MRC) expresses V-H^+-ATPase (V) on its basolateral membrane and draws Cl^- into the cell via pendrin (PDN). The pathway for basolateral Cl^- movement is unknown.*

Inhibitors of this anionic exchange reduce Cl^- uptake and produce metabolic alkalosis in freshwater fishes, and polyclonal antibodies raised against rainbow trout AE1 (anion exchanger, isoform 1) localised to the apical surface of gill cells in tilapia and coho salmon that also were immunopositive for Na^+-K^+-ATPase. In addition, an oligonucleotide probe, complementary to rat AE1 cDNA, hybridised to RNA in cells in both the filament and lamellae of the rainbow trout, suggesting that both PVCs and MRCs contain the message for AE1.

Recently, an alternative mode for Cl^-/HCO_3^- exchange across the apical membrane of the elasmobranch gill epithelium has been proposed. With the use of a heterologous antibody to human pendrin (a non-AE1 anion exchanger), pendrin-like immunoreactivity has been localised to the apical region of V-H-ATPase-rich, but not Na^+-K^+-ATPase-rich, cells in the branchial epithelium of the Atlantic stingray and spiny dogfish. This colocalisation is remarkably similar to V-H-ATPase-rich, HCO_3^--secreting intercalated cells in the mammalian collecting duct.

Intracellular CA

The branchial epithelium also contains CA, which is generally thought to be necessary for the production of H^+ and HCO_3^- for the apical exchangers, pumps, and channels that are present. Histological studies (using the cobalt phosphate/cobalt sulfate vital stain) demonstrated the presence of CA in the apical region of both MRCs and PVCs in the branchial epithelia of a variety of fish species and immunohistochemical studies (using homologous antibodies raised against purified enzyme) have confirmed this localisation to the apical region of MRCs and PVCs in both the rainbow trout and European flounder. A CAII-like gene has been sequenced from cDNA from the gills of the Osorezan dace, and the product has been localised to the MRC by homologous immunohistochemistry. In both teleosts and elasmobranchs, this intracellular CA represents the majority of CA activity in gill homogenates. A role for CA in at least Na^+uptake had been proposed, because injection of acetazolamiode into either the rainbow trout or small spotted dogfish shark inhibited gill Na^+ uptake, and it has been demonstrated that exposure of zebrafish to ethoxzolamide (another CA inhibitor) reduced both Na^+ and Cl^- uptake.

Basolateral Na^+ and Cl^- Exit

As indicated above, there is clear evidence that Na^+-K^+-ATPase is expressed in MRCs in the branchial epithelium of freshwater fishes or euryhaline (wide salinity tolerance) fishes in fresh water (teleosts and elasmobranchs). In addition, immunocytochemical studies of Na^+-K^+-ATPase (using heterologous antibodies) localised expression of the transporter to basolateral and/or tubulovesicular components of the MRCs. Partial and full-length Na^+-K^+-ATPase á-subunits have been cloned from a variety of freshwater species, and many of these studies have identified multiple isoforms. The basolateral expression of Na^+-K^+-ATPase is assumed to provide an exit step for the Na^+ from the branchial epithelial cell into the extracellular fluids. There is some evidence that one form of a Na^+-HCO_3^- cotransporter (NBC1) may be expressed in a subpopulation of MRC (defined by Na^+-K^+-ATPase immunolocalisation) in the acid-stressed Osorezan dace and the rainbow trout which could mediate both Na^+ and HCO_3^- exit to the blood. The basolateral exit step for Cl^- has not been studied directly, but an antibody raised against human CFTR localised to the basolateral regions of both MRC and PVCs in the freshwater killifish opercular epithelium, suggesting that this Cl^- channel may be involved.

Divalent Ion Uptake

Because freshwater concentrations of divalent ions such as Ca^{2+}, Mg^{2+}, and $Zn^{2+}$are far below plasma levels, fish must extract the needed ions from either food or freshwater itself. Fish (at least teleost) bone is acellular, so it does not provide a pool for Ca^{2+} as it does in terrestrial vertebrates. Recent evidence suggests that branchial uptake mechanisms exist for all three ions, but Ca^{2+} is the best studied.

Because the intracellular Ca^{2+} concentration of gill cells is presumably <1 μM, it is assumed that apical uptake is diffusional even from soft fresh water (10 μM Ca^{2+}), and partial or full-length sequences for a putative ECaC have now been reported from tilapia (B.-K. Liao, C.-H. Yang, T.-C. Pan, and P.-P. Hwang, unpublished data), fugu (GenBank accession) and rainbow trout. In both species, the putative ECaC is expressed in the gill, and in tilapia the expression increased fourfold when the fish were acclimated to low-Ca^{2+} fresh water, a condition where Ca^{2+} uptake increased (Liao et al., unpublished data). Presumed electrochemical gradients at the basolateral membrane suggest that active transport mechanisms must be involved for Ca^{2+}exit to the blood. Ca^{2+} transport across isolated membrane vesicles and the killifish opercular epithelium is Na^+ dependent, so Na^+/Ca^{2+} exchange may be involved. The complete sequence for a putative Na^+/ Ca^{2+} exchanger (NCX) has been reported from tilapia (AY283779), but its expression in the gill of this species did not change under low Ca^{2+} conditions (T.-C. Pan, C.-H. Yang, C.-J. Huang, and P.-P. Hwang, unpublished data). Ca^{2+}-dependent ATP hydrolysis can also be measured in gill vesicle preparations, with a $K_{1/2}$ of 63 nM, so a high-affinity Ca^{2+}-activated ATPase is assumed to play a role. A partial sequence for a putative Ca^{2+}-activated ATPase (PMCA) has been reported (AF236669), but its expression in the gill did not change under low Ca^{2+} conditions (Liao et al., unpublished data). Taken together, these results suggest that apical uptake, rather than basolateral extrusion, is the limiting factor in Ca^{2+} balance, at least in tilapia. The fact that Ca^{2+}-activated ATPase activity covaries with Na^+-K^+-ATPase activity in the European eel and rainbow trout, both are stimulated by prolactin injection in the eel (224), and Ca^{2+} uptake is proportional to the density of MRC in tilapia, rainbow trout, and killifish suggest that MRCs are the site of Ca^{2+} uptake by these putative transport processes.

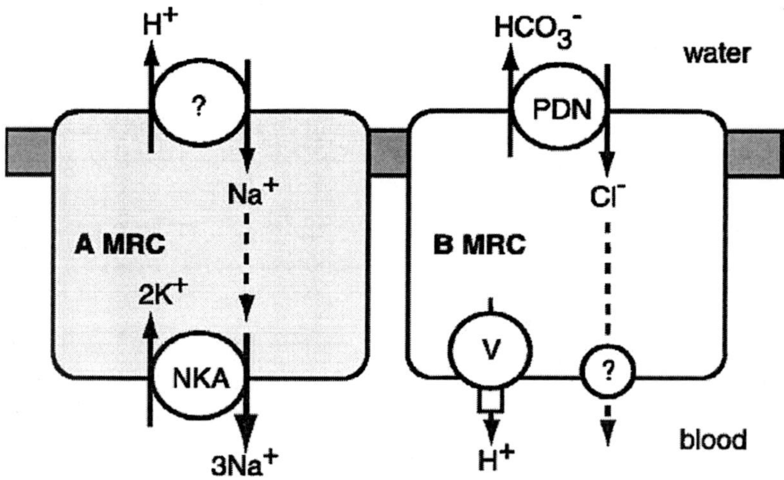

Figure: *Working model for the Ca^{2+}uptake mechanisms across the freshwater teleost gill epithelium. Ca^{2+} is drawn in via a Ca^{2+} channel (ECaC) on the apical surface of the MRC and extruded across the basolateral membrane via a Ca^{2+}-ATPase (PMCA) or a Na^+/ Ca^{2+}exchanger (NCE).*

Water Permeability and Aquaporins

Early isotopic and physiological measurements of diffusional and osmotic water permeabilities suggested that both were higher in the branchial epithelium of freshwater than that of seawater fishes, and the substantial differential between the osmotic and diffusional permeabilities suggested the presence of "water-filled pores" in the freshwater, but not seawater, gill epithelium. Aquaporin-3 (AQP-3) now has been cloned from the gills of the eel and found to be 67–69% homologous to AQP-3 from rat, mouse, and human. The expression of AQP-3 declined significantly when eels were acclimated to seawater, and a homologous antibody localised AQP-3 immunoreactivity to both the apical and basolateral aspects of branchial MRC, as well as other branchial cells (PVCs not specified). AQP-3 also has been cloned from the gills of the acid-stressed, Osorezan dace, and homologous antibodies localised expression to the basolateral aspect of the MRC. What role AQP-3 plays in osmoregulation is unknown, although it may interact with CFTR. Moreover, another aquaporin (AQP-1) has been shown to transport both ammonia and CO_2, both of which traverse the fish branchial epithelium.

Lampreys

There is little known about the specifics of gill ion transport by lampreys in fresh water, but their plasma is hypotonic to fresh water, and more than one type of MRC is present in the gill epithelium of various lamprey species when acclimated to freshwater. A recent study has demonstrated two types of MRC in the Australian pouched lamprey: one that expresses Na^+-K^+-ATPase on the basolateral aspect and another that expresses CAII in the apical area and V-ATPase in the cytoplasm. The authors propose a model for parallel pathways for Na^+ and Cl^- uptake identical to that proposed for the elasmobranch gills and similar to that found in the proximal tubule cells (for Na^+ uptake) and the type B intercalated cells (for Cl^- uptake) of the mammalian kidney. These data are all consistent with the generally accepted hypothesis that these primitive agnathan fishes possess the same gill epithelial ionic transport mechanisms as those described for freshwater teleosts, suggesting that the origin of these pathways was of the order of 500 million years ago. For more general discussions of lamprey osmoregulation in fresh water.

NaCl Secretion in Marine Teleosts

Net salt extrusion by the fish gill epithelium was first described over 70 years ago in a classic paper on Cl^- transport across the perfused gills of the eel by Ancel Keys, in which he stated: "These experiments demonstrated beyond doubt that a concentrated chloride solution is secreted by the gills in opposition to a large concentration gradient." In a subsequent paper, Keys stated that "chloride secretion exhibited by the gills of the eel in seawater is an active process requiring the expenditure of large amounts of energy," surely one of the earliest suggestions that an epithelium could actively transportions. This latter paper is also a classic in the field, because it was the first description of a specific cell involved in active transport: the "Cl^- secreting cell," which is now known to be a type of teleost MRC.

The mechanisms of NaCl secretion by at least the teleost gill epithelium are more clear than the mechanisms of NaCl uptake by the same tissue. As the importance of Na^+-K^+-ATPase in a variety of cells and epithelia (including thefish gills) became known 35 years ago, it was proposed that apical Na^+-K^+-ATPase-mediated Na^+ extrusion across the branchial epithelium, because, for example, removal of external K^+ or external application of ouabain inhibited Na^+ efflux, measured radioisotopically.

This proposition of apical Na^+-K^+-ATPase extruding cellular Na^+ was rendered untenable when Karnaky demonstrated unequivocally that Na^+-K^+-ATPase was restricted to the basolateral membrane of the MRC in the branchial epithelium of the killifish, and this localisation has now been confirmed for a variety of species, including elasmobranchs. The importance of basolateral Na^+-K^+-ATPase in both Na^+ and Cl^- extrusion was demonstrated by the fact that injection of ouabain into the eel completely inhibited both Na^+-K^+-ATPase activity and the efflux of both ions.

It was proposed that NaCl extrusion was mediated by a basolateral cotransport of NaCl down the electrochemical gradient produced by Na^+-K^+-ATPase, coupled with an apical extrusion of Cl^- via a channel and paracellular extrusion of Na^+, both down their respective electrochemical gradients. This model was consistent with work on other epithelia and was corroborated by concurrent, electrophysiological studies of the epithelial sheets from the operculum of the killifish and the lower jaw of the longjaw mudsucker.

The killifish opercular epithelium contained ~60% MRCs, maintained an open-circuit electrical potential of 19 mV (serosal side positive) and an electrical resistance of 174 Ù · cm^2, and produced a net short-circuit current (I_{sc}; 137 $\mu A/cm^2$) that was equivalent to the net Cl^- efflux, measured isotopically. Addition of ouabain or furosemide, or ionic substitutions to the basolateral side, inhibited the I_{sc} and effluxes of Na^+ and Cl^-, consistent with the model of basolateral Na^+-K^+-ATPase and NaCl cotransport. Studies showing a direct correlation between MRC number and Cl^- currents in the opercular and jaw skin epithelia, combined with salinity-induced changes in structure and number of MRCs in fish gills suggested the primacy of the MRC in

NaCl extrusion. This was proven directly when Cl^- currents were localised to the MRC of the killifish operculum, using a vibrating probe technique.Below diagrams the generally accepted model for NaCl extrusion by the teleost MRC, and supporting evidence for specific pathways follows.

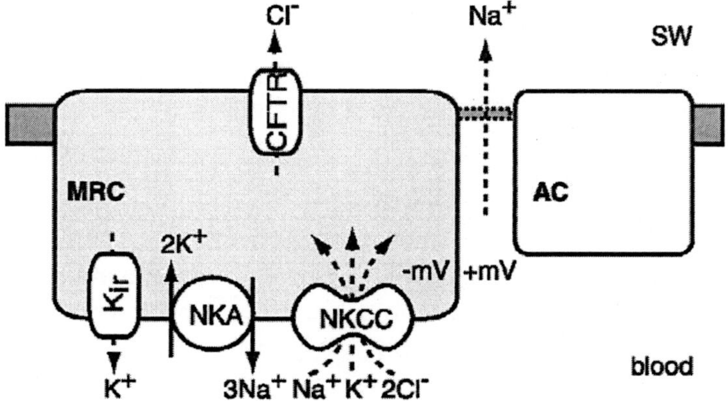

SW Teleosts

Figure: *Working model for the extrusion of NaCl by the marine teleost gill epithelium. Plasma* Na^+*,* K^+*, and* Cl^- *enter the cell via basolateral NKCC;* Na^+ *is recycled back to the plasma via* Na^+*-*K^+*-ATPase and* K^+ *via a* K^+ *channel (*K_{ir}*).* Cl^- *is extruded across the apical membrane via a* Cl^- *channel (CFTR). The transepithelial electrical potential across the gill epithelium (plasma positive to seawater) drives* Na^+*across the leaky tight junctions between the MRC and the AC.*

Basolateral Na^+ and Cl^- Uptake

NA^+-K^+-ATPASE

Complete or partial sequences for Na $^+$-K^+-ATPase now have been reported from a variety of freshwater and euryhaline teleosts as well as stenohaline marine species from the Antarctic and New Zealand, where multiple isoforms were identified. Na^+-K^+-ATPase expression can be localised to the MRCs in the gill epithelium in marine species (including the dogfish shark) or euryhaline species in seawater (including the stingray) using nucleotide probes and antibodies. In some cases, Na^+-K^+-ATPase has been specifically localised to the basolateral aspect of the MRC Expression and activity of Na^+-K^+-ATPase is often correlated directly with salinity oreover, relative Na^+-K^+-ATPase activity in gill biopsies can actually predict the future downstream migratory behaviour in freshwater brown trout. On the other hand, Na^+-K^+-ATPase activity and/or expression actually decreases when some euryhaline species are acclimated to seawater. The expression of Na^+-K^+-ATPase increased (compared with seawater controls) when the sea bass (*Dicentrarchus labrax*) is acclimated to either fresh water or 200% seawater, which suggests that any osmotic stress might affect the expression of this important transport enzyme. Some of these discrepancies may be due to differential expression of Na^+-K^+-ATPase isoforms, because a recent study has demonstrated that, in the rainbow trout gills, Na^+-K^+-ATPase a1b is upregulated in seawater, but Na^+-K^+-ATPase á1a is downregulated.

Apical Salt Extrusion

Marshall's group has identified a low-conductance (8 pS) anion channel in a primary culture of cells from the killifish opercular epithelium (which retained MRC) that was stimulated by cAMP and inhibited by two anion channel blockers, diphenyl-amine-2-carboxylic acid (DPC) and 5-nitro-2-(3-phenylpropylamino)benzoic acid (NPPB). A full-length clone of a CFTR gene was sequenced from a killifish gill cDNA library that encoded a protein (59% identical to human CFTR amino acid sequence) and that produced a cAMP-activated anion conductance when its mRNA was expressed in *Xenopus* oocytes. Importantly, expression of the product in the gills increased severalfold after transfer of killifish to seawater.

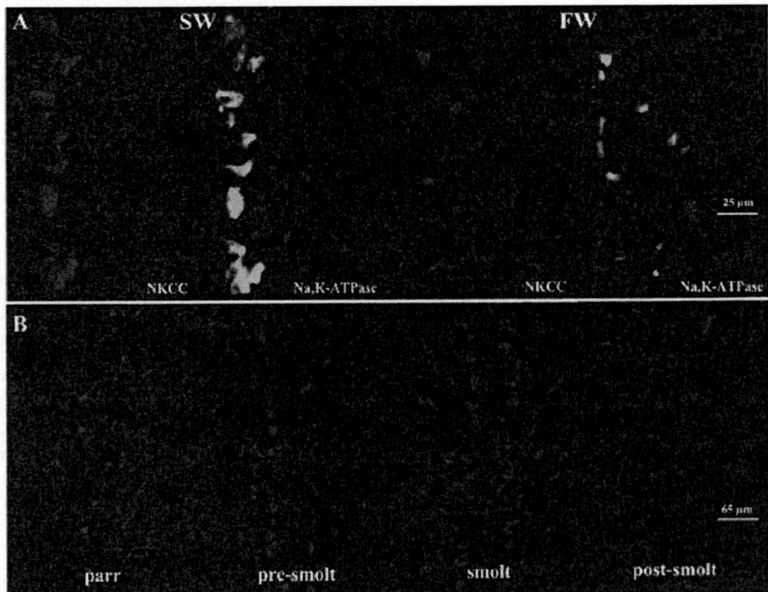

Figure: *Confocal laser scanning images of kfCFTR immunofluorescence of mouse anti-human CFTR and goat anti-mouse IgG Oregon Green 488*

in killifishopercular epithelium show the distribution of kfCFTR (green fluorescence + red yields yellow) in mitochondria-rich (MR) chloride cells counterstained for mitochondria with Mitotracker Red. A*: membrane from a SW-adapted animal has kfCFTR immunofluorescence at 4.5 μm from the surface of the chloride cells is localised to the apical crypt (white arrows). The asterisk indicates a gap in the pavement cell layer (*B*). The same frame as for* A *but at a depth of 7.5 μm at the plane of the nucleus of the MR cell. Blue arrows in* B *indicate the mitochondria-rich chloride cells below the apical crypt.* D *and* E*: similar kfCFTR immunostaining, but in MR cells from fish acclimated to FW. An optical section (*D*) at 4.5 μm from the surface has positive kfCFTR immunofluorescence evenly distributed in the chloride cells (white arrows) as well as in the pavement cells at the plane of the pavement cell nuclei (orange arrow). Optical section (*E*) is same frame as for* D *but 10.5 μm into the tissue at the level of the MR cell nuclei. Blue arrows indicate the MR cells with diffusely distributed positive immunofluorescence for kfCFTR.* F *and* G*: similar immunostaining but for a fish transferred from FW to SW for 48 h.* F*: an optical section 9.0 μm from the surface with ring-shaped kfCFTR positive immunofluorescence (white arrows).* G*: 13.5 μm from the surface with kfCFTR immunofluorescence diffusely distributed in MR cells (blue arrows). Scale bars in* A, D, *and* F *are 20 μm.*

Full-length clones of CFTR are now available from the Atlantic salmon (95% identical to killifish amino acid sequence) and fugu (*Fugu rubripes*, 84% identical to killifish amino acid sequence).

Antibodies directed against hCFTR have localised expression to the apical region of MRC in the giant mudskipper and Hawaiian goby, and a monoclonal antibody directed against the shared carboxy terminus of hCFTR and kCFTR confirmed this location in the killifish gill epithelium. In the latter study, the immunoreactive CFTR moved from the central part of the cytosol to the apical surface of the MRC when the killifish were acclimated to seawater.

In the Hawaiian goby, CFTR-positive cells increased fivefold in seawater-acclimated fish. For an excellent review of the role of CFTR in the seawater teleost gill.

Does the Elasmobranch Gill Excrete NaCl?

The fact that removal of the rectal gland from at least *Squalus acanthias* does not result in significant hypernatremia suggests that other tissues can secrete salt when necessary.

The elasmobranch gill epithelium does contain MRCs, which express immunoreactive Na^+-K^+-ATPase, and NKCC expression has been noted in the dogfish gills.

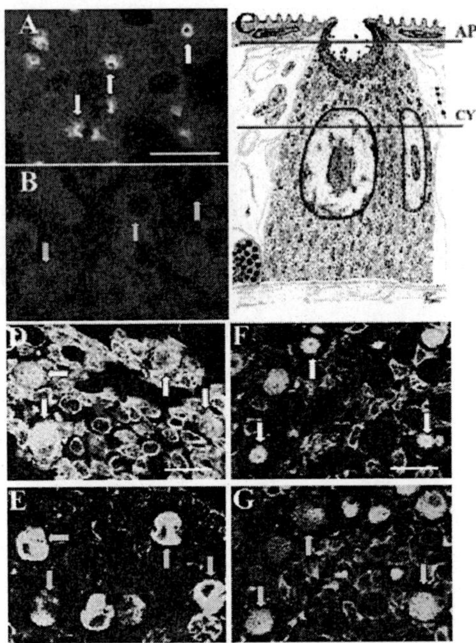

Figure: *Confocal laser scanning images of kfCFTR immunofluorescence of mouse anti-human CFTR and goat anti-mouse IgG Oregon Green 488 in killifishopercular epithelium show the distribution of kfCFTR (green fluorescence + red yields yellow) in mitochondria-rich (MR) chloride cells counterstained for mitochondria with Mitotracker Red.* A*: membrane from a SW-adapted animal has kfCFTR immunofluorescence at 4.5 μm from the surface of the chloride cells is localized to the apical crypt (white arrows). The asterisk indicates a gap in the pavement cell layer (*B*). The same frame as for* A *but at a depth of 7.5 μm at the plane of the nucleus of the MR cell (see drawing,* C*, for approximate depths of the optical sections; AP, apical crypt level; CY, cytosol level). Blue arrows in* B *indicate the mitochondria-rich chloride cells below the apical crypt.* D *and* E*: similar kfCFTR immunostaining, but in MR cells from fish acclimated to FW. An optical section (*D*) at 4.5 μm from the surface has positive kfCFTR immunofluorescence evenly distributed in the chloride cells (white arrows) as well as in the pavement cells at the plane of the pavement cell nuclei (orange arrow). Optical section (*E*) is same frame as for* D *but 10.5 μm into the tissue at the level of the MR cell nuclei. Blue arrows indicate the MR cells with diffusely distributed positive immunofluorescence for kfCFTR.* F *and* G*: similar immunostaining but for a fish transferred from FW to SW for 48 h.* F*: an optical section 9.0 μm from the surface with ring-shaped kfCFTR positive immunofluorescence (white arrows).* G*: 13.5 μm from the surface with kfCFTR immunofluorescence diffusely distributed in MR cells (blue arrows). Scale bars in* A, D, *and* F *are 20 μm.*

However, neither MRC number or size nor Na^+-K^+-ATPase activity increased after the removal of the rectal gland in the dogfish, counter to what one would expect if branchial extrusion mechanisms were upregulated in response to the loss of the rectal gland.

Moreover, acclimation of freshwater Atlantic stingrays to seawater was associated with a decrease in branchial Na^+-K^+-ATPase activity and Na^+-K^+-ATPase immunoreactivity. More studies are warranted, but it is possible that increased renal Na^+ and Cl^- excretion could offset diffusional salt uptake in the absence of the rectal gland, because the elasmobranch gains water by osmosis and, therefore, has high urine flow rates.

As long as the rate of water absorption via osmosis is greater than or equal to diffusional salt uptake, blood osmolarity can be maintained by excretion of urine that is isotonic or even hypotonic to the plasma. Indeed, Burger found increased urinary salt loss after removal of the rectal gland in the spiny dogfish.

Hagfish and Lampreys

Despite the lack of significant osmotic or NaCl gradients across their gills, the hagfish branchial epithelium does contain cells that share many of the morphological characteristics that define the MRC of teleosts and elasmobranchs. Moreover, Na^+-K^+-ATPase can be immunolocalised to MRC on the filaments and lamellae of the gill epithelium of the Atlantic hagfish, and NHE-1 expression was detected in gill extracts by Western blots in the same species. Intracellular distribution was not determined, but expression of these two proteins suggests that the MRC in the gills of the hagfish may be involved in acid-base regulation, as it is in teleosts and elasmobranchs.

All lampreys breed in fresh water, and one group, the petromyzonitids, enter seawater after metamorphosis. The marine lampreys (e.g., *P. marinus*) migrate back to fresh water as breeding adults, and one population is actually resident in the Great Lakes of North America. Little is known about osmoregulation in the marine populations because of the difficulty of capturing animals, but MRCs in the gill epithelium of marine lampreys lose basolateral invaginations after acclimation to fresh water, and the activity of gill Na^+-K^+-ATPase is much higher in seawater versus freshwater individuals, suggesting a role for Na^+-K^+-ATPase in MRCs in seawater osmoregulation. More data are needed, but it is generally assumed that the gill mechanisms for osmoregulation in seawater lampreys parallel those in the marine teleost gills. Modern molecular techniques should allow confirmation of this assumption.

pH Regulation

Fishes, like all vertebrates, have three compensatory mechanisms to regulate the acid-base status of their extracellular body fluids: *1*) instantaneous physiochemical buffering with bicarbonate and nonbicarbonate buffers, *2*) respiratory adjustments of the "open" CO_2-HCO_3^- buffer system within minutes, and *3*) net transport of acid-base relevant molecules between the animal and environment within minutes to hours. Using an aqueous medium for respiration limits the first two mechanisms relative to air-breathing vertebrates, but enhances the third mechanism.

Respiratory Compensation

In well-aerated water, the obligatorily high gill ventilation rates and high capacitance of CO_2 allows metabolically generated CO_2 to readily leave through the gill without a substantial buildup in the tissues of fishes, as it may in air-breathing vertebrates. This results in an arterial PCO_2 that is never more than a few mmHg above the water irrigating the gills (Pa_{CO2} ~1 mmHg in normocapnic water), a minute CO_2 gradient compared with air-breathing vertebrates such as mammals. Accordingly, the blood bicarbonate concentration of fishes is also low (~4 mM in normocapnic water compared with 24 mM in humans)The low steady-state Pa_{CO2} and bicarbonate concentration minimise the ability of fishes to "blow-off" buffered metabolic acid loads via hyperventilation as occurs in mammals and reduces the buffering capacity of the extracellular compartment during metabolic acid-base disturbances.

For example, data on longhorn sculpin, lemon sole (*Parophrys vetulus*), and rainbow trout show that acute mineral acid infusions actually cause a slight respiratory acidosis (in addition to a severe metabolic acidosis), suggesting a lack of any respiratory compensation. The same was shown for several species when lacticacidosis was induced by strenuous exercise; a slight respiratory acidosis (Pa_{CO2} increase of 2–3 mmHg) developed that lasted for 3–10 h even though hyperventilation occurred in response to internal hypoxia.

Although respiratory compensation for metabolic acidosis does not appear to exist in fishes, several elasmobranch and teleost species have been shown to hyperventilate in response to environmental hypercapnia.

However, this mechanism is not an effective compensation for respiratory acidosis for three reasons: *1*) it only responds to hypercapnia of external origin, and not to hypercapnia of endogenous origin; *2*) even if hyperventilation was pronounced enough to abolish the Pv_{CO2} gradient between blood and environmental water completely, the blood pH would only be raised by one- or two-tenths of a pH unit (calculated from Henderson-Hasselbalch equations) above what it would be in the absence of hyperventilation because of the small steady-state PCO_2 gradient at the gills; and *3*) hyperventilation can never fully compensate for an environmental hypercapnia, because the source of the acidosis is the respiratory medium. Therefore, the PCO_2-sensitive hyperventilation is more likely to have evolved as a "back-up" of PO_2-sensitive hyperventilation to maintain O_2 uptake at the gills, because hypoxia and hypercapnia usually occur simultaneously in natural waters.

Acid Secretion

V-ATPase

Pharmacological agents have been used to determine the possible mechanisms of acid secretion by the gills of several fish species. In addition to blocking Na^+absorption, amiloride has been shown to inhibit net-acid excretion when placed in water irrigating the gills of rainbow trout, brown trout, European flounder, killifish, little skate, and longhorn sculpin. However, these results do not distinguish between the two acid secretion mechanisms, because amiloride inhibits Na^+ transport by NHEs and ENaC.

For example, amiloride would be expected to inhibit acid secretion from an NHE by blocking the exchanger directly, but could also theoretically inhibit acid secretion by inhibiting an ENaC-like channel and thereby altering the apical electrochemical gradient. However, three studies were able to eliminate NHE as the predominant acid transporter in three freshwater teleost species.

In rainbow trout, 0.1 mM amiloride had no effect on net acid excretion, a concentration that inhibits >80% of Na^+ absorption in this species. Similarly, 1 mM amiloride had no effect on net acid excretion from brown trout but did inhibit ~70% of Na^+ absorption. Finally, 5 mM 5-(*N*,*N*-dimethyl)amiloride (HMA), a specific inhibitor of NHEs, had no effect on net acid excretion (or Na^+ absorption) from isolated gill filaments of freshwater European flounder. These demonstrations of decoupled acid secretion and Na^+ absorption are strong arguments against a direct role for apical NHEs in freshwater teleosts, and together with the apical localisation of V-ATPase in rainbow trout, coho salmon, and tilapia, help support the role of V-ATPase as the predominant mechanism of acid secretion in freshwater teleosts.

Several studies have attempted to determine if the quantity of V-ATPase expression in freshwater teleost gills is increased during acidosis, which would be expected if this proton pump were responsible for the majority of acid secretion. In support of this hypothesis, a recent study used real-time PCR (RT-PCR) on rainbow trout gills to demonstrate large, sustained increases (4- to 70-fold increases) in V-ATPase expression during intravascular infusions of HCl. Unfortunately, the results for hypercapnia are conflicting, even though almost all of them were conducted on a single family of fishes (Salmonidae, mainly rainbow trout). For example, Northern blots and RT-PCR detected a transient 1.5- to 2.0-fold increase in B subunit expression in the first 2 h of chronic hypercapnia, but this increase subsided to control levels by 4 h and was never again elevated for the rest of the experiment (24 h total).

Other studies, using the same species and subunit, suggested increased mRNA and protein expression after 18 h of hypercapnia, a time frame that is out of phase with the RT-PCR and Northern blot results. In contrast, a preliminary study on rainbow trout failed to detect any change in A subunit protein levels after 40 h of hypercapnia (405), and a study on another salmonid (Atlantic salmon) actually measured a small, but consistent, decrease in B subunit mRNA expression after 24 h of 2% hypercapnia.

Unfortunately, the reasons for these inconsistent results are unknown, but they do suggest that large changes in the expression of V-ATPase probably are not an important mechanism of regulating net acid secretion during hypercapnia. Interestingly, analogous tissues (such as mammalian renal tubule collecting ducts) appear to use rapid changes in the location (vesicular vs. apical) and not the total expression of V-ATPase pumps, as mechanisms of regulating acid secretion.

Prior to studies that localised V-ATPase and measured changes in the enzyme's expression, a series of ultrastructural studies were conducted to determine if morphological changes to PVC and/or MRC occur in response to acid-base disturbances, as they do in mammalian renal tubule intercalated cells.

Scanning electron micrographs demonstrated a 30 and 85% decrease in apical MRC fractional area (MRCFA) during hypercapnia in rainbow trout and brown bullhead catfish, respectively. They were also used to measure a 50% increase in MRCFA during metabolic alkalosis in rainbow trout. These changes in MRCFA were directly proportional to changes in unidirectional Cl^- uptake and inversely proportional to changes in net acid secretion and unidirectional Na^+ uptake, suggesting that MRCs are responsible for apical Cl^-/HCO_3^- exchanges and not for acid secretion or Na^+absorption. Alternatively, a 100% increase in the density of PVC microvilli was measured in response to hypercapnia in brown bullhead catfish, suggestive of an increase in apical cell area of this cell type. It was proposed that net acid secretion was increased by removal of Cl^-/HCO_3^-exchangers from the apical membrane of MRC and/or insertion of proton pumps into the apical membrane of PVCs, a model that is consistent with the expression of apical Cl^-/HCO_3^- exchangers in MRCs and V-ATPase in PVCs.

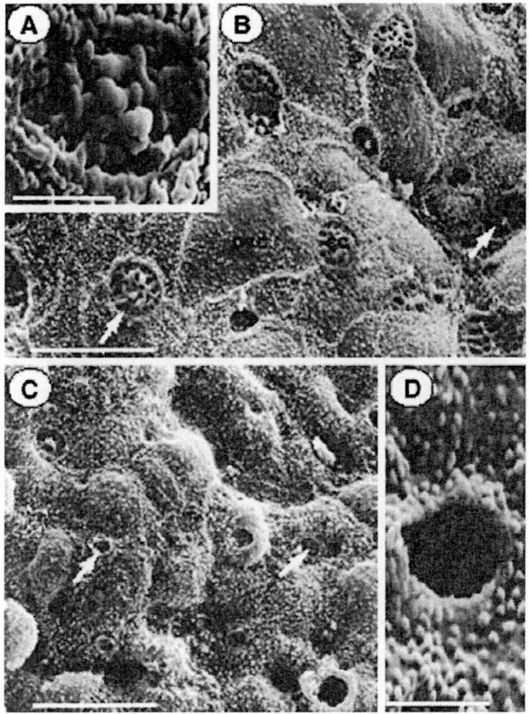

Figure: *Representative low- and high-magnification scanning electron micrographs of the filamental epithelium from brown bullhead catfish (*Ictalurus nebulosus*) during normocapnia (*A *and* B*) and after 6 h of environmental hypercapnia (*C *and* D*). Hypercapnia caused a reduction in apical mitochondrion-rich cell fractional area (i.e., MRC fractional area) and a change in the appearance of the MRCs.* A *and* D *are high magnifications of the cells marked with white arrows in* B *and* C*, respectively. Bars: 2 μm in* A *and* D*; 20 μm in* B *and* C.

However, one ultrastructural result contradicted this model: a metabolic acidosis was correlated with a 135% increase in MRCFA of rainbow trout. One explanation raised for this conflicting result was the possibility of two or more MRC subpopulations, one with apical Cl^-/HCO_3^- exchangers and one with apical proton pumps, which is consistent with studies that have localised V-ATPase in MRCs and PVCs.

Therefore, although more work is needed to confirm the specific roles of MRCs and PVCs, opposing changes in the location of acid and base transporters (via altered apical membrane morphologies) may be important mechanisms for regulating rates of net acid excretion from the gills of freshwater teleosts.

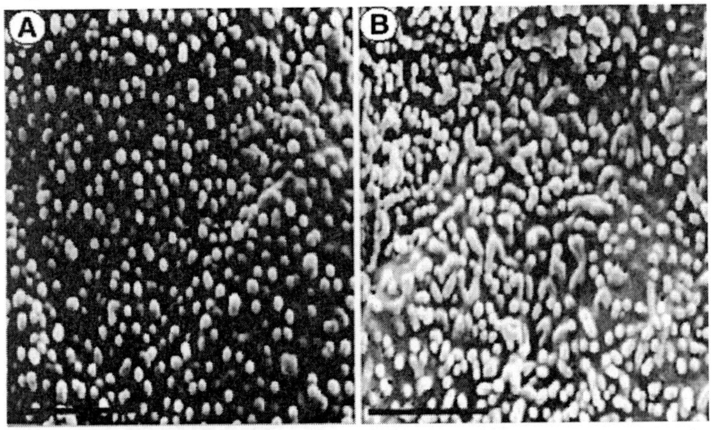

Figure: *Representative high-magnification scanning electron micrographs of pavement cells (PVCs) on the filamental epithelium of brown bullhead catfish (*Ictalurus nebulosus*) during normocapnia (*A*) and after 6 h of environmental hypercapnia (*B*). Hypercapnia caused an increase in microvilli density. Bars: 2 μm.*

It is important to note that with the exception of tilapia, all of the immunological evidence for apical V-ATPase is limited to salmonids, and therefore, it is premature to propose that V-ATPase is the sole route of acid secretion in all freshwater fishes. Moreover, with the exception of European flounder, all of the evidence against a direct coupling between Na^+ and H^+ are from salmonids, and therefore, it is premature to propose that apical NHEs do not function in any freshwater fishes.

In fact, recent studies on killifish, Osorezan dace, and Atlantic stingrays suggest that V-ATPase is not an important route of apical acid secretion in all fishes. Moreover, a homologous antibody for V-

ATPase was used to exclusively localise the proton pump to the basolateral membranes of Na^+-K^+-ATPase-rich MRCs in the gills of freshwater acclimated killifish. Although mechanisms of acid-base transport were not proposed by the authors, it is clear that V-ATPase cannot be the direct route of acid secretion from this species. Previous pharmacological studies on killifish demonstrated a strong link between Na^+ absorption and net acid excretion when 0.1 mM amiloride was applied, which is more consistent with an apical NHE than V-ATPase and an ENaC-like channel.

In a molecular and immunological study by Hirata et al., control Osorezan dace were shown to have low levels of V-ATPase B subunit expression, which increased only slightly when exposed to low pH water (3.5) for up to 5 days. In light of much larger increases in other acid transporters, it was suggested that V-ATPase does not play a large role in acclimation to low pH (a condition that causes metabolic acidosis), and an alternative model was presented.

Finally, V-ATPase may have roles in base secretion and Cl^- absorption, as opposed to acid secretion, in nonteleost fishes. As mentioned in section V*B1*, V-ATPase has been localised to the basolateral region of MRCs that contain apical immunoreactivity for pendrin in fresh and seawater elasmobranchs. Similarly, V-ATPase is basolateral in type B intercalated cells of mammalian renal tubule collecting ducts and therefore provides a route of acid exit from the serosal side of the cells, in addition to generating base for pendrin (the apical base secreting anion exchanger). This function is also presumed for V-ATPase and pendrin-rich cells of elasmobranchs.

NHE

Intuitively, NHEs might be proposed as the dominant route of acid secretion fromfishes in seawater, where the high Na^+ concentration favours Na^+ entry across the apical membrane. HMA (0.1 mM) was shown to inhibit 75% of net acid excretion from longhorn sculpin in brackish water, suggesting that NHEs are responsible for the majority of acid secretion from this seawater teleost. This result and the finding of reduced or absent immunoreactivity for V-ATPase in seawater rainbow trout and coho salmon have led the general idea that NHEs are the proteins responsible for acid secretion at the apical side of seawater fish gills. Unfortunately, the effects of specific NHE inhibitors on acid secretion from seawater fishes have only been reported for longhorn sculpin, so it is not known if this generality can be applied to all seawater fishes.

Despite the relatively few pharmacological data that support a direct role in acid excretion, NHEs are clearly expressed in the gills of many fishes, and recent studies have demonstrated large increases in expression during treatments that cause acidosis. For example, full-length transcripts homologous to NHE2 have been cloned from longhorn sculpin, killifish, and spiny dogfish, and a full-length transcript homologous to NHE3 has been reported from freshwater Osorezan dace.

In addition to their previously mentioned roles in Na^+ absorption, these isoforms are responsible for acid secretion from mammalian intestinal and renal proximal tubule cells, and therefore are predicted to facilitate Na^+/H^+ exchange at the apical membrane of seawater fishes. Before the availability of these full-length fish sequences, heterologous antibodies were used to localise NHE2 and -3 immunoreactivity in many species of seawater teleosts and elasmobranchs. In general, the immunoreactivity was located in MRCs, and in some cases colocalised with Na^+-K^+-ATPase. This is consistent with models that predict that basolateral Na^+-K^+-ATPase creates a low intracellular [Na^+] to drive apical Na^+/H^+ exchange.

On the other hand, a homologous NHE antibody was made recently for NHE2 of the spiny dogfish, and NHE2 immunoreactivity was localised to a population of MRCs that were not rich in Na^+-K^+-ATPase. The membrane location of this shark NHE2 was not obvious in light or confocal micrographs, so its function(s) in this cell type remain(s) to be determined. Future immunohistochemcal studies with homologous antibodies to the recently sequenced NHEs will help determine if these localisation discrepancies are due to isoform, antibody, and/or species differences.

The gill of hagfish also has been shown to express an NHE, but a 1,047-bp transcript was found to be only 38–50% homologous to NHEs in GenBank. Despite this low homology, a hydropathy plot of the hagfish NHE suggested a similar topology to mammalian NHEs. In addition, relative RT-PCR was used to demonstrate an increase in expression following an injection of H_2SO_4, a treatment that causes metabolic acidosis. The increased expression was highest 2 h after the injection, a time that is consistent with the maximum rates of net-acid excretion, suggesting a direct role for this putative agnathan NHE in systemic acid-base regulation.

A recent study used homologous antibodies and molecular probes to support the hypothesis that NHE3 is the route of acid secretion from a unique freshwater teleost, the Osorezan dace. This is a freshwater teleost that lives in the extremely acidic (pH 3.4–3.8) Lake Osorezan of Japan. In this acid-tolerant fish, NHE3 immunoreactivity was localised to the apical side of MRCs that also expressed CAII, Na^+-K^+-ATPase, and NBC1.

Northern blots were also used to demonstrate that the mRNA expression of NHE and other transport related enzymes (CAII and NBC1) increased greatly when dace were transferred from water of neutral to highly acidic pH (3.5), a condition that causes metabolic acidosis. Interestingly, NHE3 is responsible for most of the acid secretion from brush-border membranes of mammalian renal proximal tubule cells where its expression also increases during metabolic acidosis. Mammalian renal proximal tubules also express CAII, Na^+-K^+-ATPase, and NBC1, making a compelling argument that the dace NHE3-rich cells function in a similar manner.

The most interesting characteristic of this study is that it suggests that an NHE-dependent mechanism may be functioning in fresh waters with a Na^+ concentration of ~1 mM and a pH below 4.0, concentrations that should make an apical NHE thermodynamically unfavourable. Although CAII, Na^+-K^+-ATPase, and NBC1 activity in the same cells would be expected to help make H^+ and Na^+ gradients more favourable by lowering intracellular $[Na^+]$ and pH, it is hard to imagine how a >1,000-fold uphill H^+ gradient could be overcome. However, two other strategies might be involved: *1)* the expression of glutamine dehydrogenase (GDH) increased markedly in all tissues in acidic water, and *2)* the NHE3-rich cells formed a follicle in acidic water. GDH is a mitochondrial enzyme that catalyses NH_4^+ and HCO_3^- production from the substrate glutamine, and the increase in its expression suggests large increases in NH_4^+ and HCO_3^- synthesis in low pH water.

Presumably, the increased NH_4^+ synthesis would be secreted from the gills, leaving a net base HCO_3^- gain in the fish. In mammals, acidosis causes this to happen predominantly in the kidneys, where NH_4^+ is secreted by NHE3 in the proximal tubule. Therefore, as suggested by Hirata et al., the large proton gradient in dace may be overcome by secreting NH_4^+ via NHE3, instead of H^+; a demonstration of higher affinity for NH_4^+ than H^+ would help support this model. Alternatively, ammonia may leave the gills in its basic gas form (NH_3) by simple diffusion, as it does in most fishes. This would buffer the water adjacent to the gills and therefore decrease the proton gradient inhibiting acid secretion. The follicular structure might help by partially sealing the NHE3-rich cells from the low pH of the bulk water and preventing immediate loss of buffers from its lumen. Clearly, more studies like these are needed with fish-specific immunological and molecular probes to determine if NHE3 and/or NHE2 expression are regulated during acid-base disturbances in other fishes, and where they are located in the gills.

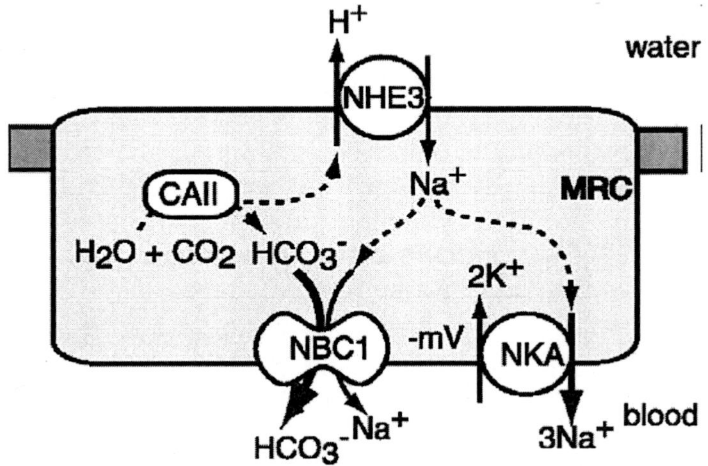

Acidic Osorezan Dace

Figure: *Model of acid secretion and Na^+ absorptive mechanisms in gill MRCs of FW Osorezan dace. Expression of transcripts for all four proteins were shown to increase during exposure to acidic water (pH 3.5) and were localised to a single cell type with dace-specific antibodies. In the model, acid secretion and Na^+ absorption are initiated by Na^+-K^+-ATPase, which produces a low intracellular [Na^+] and a negative inside membrane potential. These conditions then favour Na^+ absorption, in exchange for acid secretion through an apical NHE3, which increases the intracellular pH. The higher pH increases intracellular [HCO_3^-] via CO_2 hydration by carbonic anhydrase II. Finally, the increased intracellular [HCO_3^-] and negative potential drive electrogenic efflux of Na^+ and HCO_3^- across the basolateral membrane through NBC1. Electrogenic transport is indicated with unequal arrow weights. Solid arrows indicate facilitated transport, and broken arrows indicate diffusion.*

H^+-K^+-ATPase

Although NHEs and V-ATPase are apical acid transporters in fish gills (and most models predict that one or the other functions in a given species), it is possible that other acid transporters also may be expressed in the gill epithelium. For instance, mammalian renal epithelial cells have H^+-K^+-ATPases, as well as NHEs, V-ATPases, along their luminal membranes. These transporters are divided further into isoforms that are thought to function in different nephron segments and may be stimulated by different pH and/or electrolyte imbalances. It has recently been demonstrated that a transcript homologous to an H^+-K^+-ATPase is expressed in the gills of elasmobranchs.

RT-PCR was used with mRNA from spiny dogfish and Atlantic stingrays to obtain sequences that are over 80% identical to HKá1 sequences at the amino acid level. In addition, two antibodies made for different parts of mammalian HKá1 stain the gills of marine Atlantic stingrays. However, the subcellular location of this staining is not clear, and RT-PCR comparisons of expression between control and hypercapnic stingrays failed to detect a difference, making further expression and localisation studies necessary to understand HKá1's role in gill acid secretion.

NBC

Base efflux across the basolateral membrane must compliment acid secretion at the apical membrane for net systemic acid excretion to occur in epithelial cells. The complete cDNA sequence of a homolog of a base transporter, NBC1, recently has been reported from the gills of rainbow trout and Osorezan dace, and a partial sequence from the gill of Atlantic stingrays is also known (Choe and Evans, GenBank accession no. AY652419).

NBC1 facilitates Na^+ and HCO_3^- efflux across the basolateral membrane of mammalian proximal tubules in a 1:3 Na^+-to-HCO_3^- ratio and has been hypothesized to have a similar function in fish gills. In support of this proposed role in systemic acid excretion, the dace NBC1 homolog was localised to the basolateral membranes of Na^+-K^+-ATPase-rich MRCs and its expression increased during exposure to low pH water.

Similarly, expression of the rainbow trout gill NBC1 homolog increased during hypercapnia. As discussed by Perry et al., it is also possible that the putative fish NBC1 homologs function in an influx mode, as occurs in mammalian pancreas cells. This would alkalise cells and would still be consistent with an increased expression during acidosis as a mechanism of intracellular pH regulation. However, at least for Osorezan dace, NBC1 was specifically localised in MRCs that were also shown to have apical NHE3 and basolateral Na^+-K^+-ATPase. Because the activities of these two highly expressed transporters would alkalise and hyperpolarise the cell, it is more likely that dace

NBC1 functions in an efflux mode. Further studies are clearly needed to determine the role of NBC1 in fish gills.

Nitrogen Balance

Unlike biomolecules such as fatty acids and sugars, excess amino acids are not stored or excreted. Their carbon skeletons are instead converted into intermediate metabolic fuel sources after removing the á-amino group. While the carbon backbone is converted into fatty acids, ketone bodies, and carbohydrates, the excess nitrogen is usually excreted. This net gain in excess nitrogen is magnified in the high-protein diets of many fishes where most of their dietary carbon is extracted from amino acids.

In addition, fasting fishes can use their own muscle proteins as a source of amino acids, which are deaminated to produce ATP for maintenance, or converted to glucose by the liver. Excess nitrogen is not usually stored in tissues because it is either too toxic (i.e., ammonia) or too energetically costly to convert to less toxic forms (i.e., uric acid and urea). For example, ammonia, the most useful form of nitrogen for synthesizing amino acids, is highly toxic and therefore cannot be sequestered at any significant level. Although less toxic than ammonia, urea and uric acid require ATP consumption for their synthesis, counteracting the benefits of catabolising proteins for energy. In addition, uric acid precipitates out of solution and is produced by terrestrial animals that need to conserve water (e.g., birds and reptiles), obviously not a problem for at least freshwater fishes.

Somefishes (elasmobranchs and coelacanths) do synthesize and sequester nitrogen as urea but for osmoregulation, not nitrogen storage. In fact, eukaryotes lack ureases, and therefore chondrichthian fishes cannot reincorporate the urea they synthesize unless they harbour urease-expressing bacteria. Although an apparently successful strategy (>1,000 extant species), the high urea concentrations of ureosmotic fishes (>300 mM) are toxic and must be balanced with stabilising molecules such as trimethylamine oxide. Therefore, net production of nitrogenous waste occurs in most fishes, under most conditions, and because of toxicity and energetic costs of storage, most fishes deal with waste nitrogen by excreting it immediately. Nitrogenous molecules other than ammonia and urea are excreted by fishes (e.g., taurine, creatine, creatinine, purines, and methylamines), but their quantitative contributions are small and mechanisms of excretion poorly understood. Mechanisms of ammonia and urea metabolism and excretion have been studied extensively in fishes, and several excellent reviews are available. Here, we focus on mechanisms of excretion by the gills.

The gills are the primary site of ammonia and urea excretion from fishes for many of the same reasons that they are the primary site of gas exchange and systemic ion and acid-base transport (e.g., large surface area, perfusion by 100% of cardiac output, large ventilation rates, small diffusion distances, and contact with a voluminous mucosal medium). Branchial and renal nitrogen excretion rates have been compared in several species, and >80% of the total nitrogen (ammonia plus urea) was excreted from the gills of most species.

The vast majority of fishes, including almost all actinopterygians and agnathans, excrete the majority of their nitrogenous waste as ammonia and are referred to as ammonotelic. This appears to be the default condition for fishes, reflecting the ease with which NH_3 can diffuse through tissues and into a large external aqueous environment. Alternatively, coelacanths, most elasmobranchs, and a few teleosts excrete most of their nitrogenous waste as urea and are referred to as ureotelic. Ureotelism appears to have evolved either as a mechanism of detoxifying ammonia in environments that inhibit ammonia excretion (e.g., air exposure and high pH) or as a mechanism of osmoregulation in seawater (coelacanths and elasmobranchs).

Ammonia

Ammonia is a weak base and occurs as both a gas, ammonia (NH_3), and an ion, ammonium (NH_4^+), in aqueous solutions; the sum of both forms is referred to as total ammonia (T_{amm}). Because the pH of fish blood and intracellular fluid is over one unit below the p*K* of ammonia (~9–10), ~95% of T_{amm} exists as NH_4^+ infish tissues. Although the specific mechanisms of T_{amm} toxicity in animals are not completely understood, high micromolar concentrations are clearly incompatible with many tissue functions. The most acute effects of ammonia are probably related to the ability of its ionic form, NH_4^+, to substitute for K^+ in ion transporters and disrupt electrochemical gradients in central nervous systems. This may explain why fishes, which have less advanced central nervous systems, are more tolerant of T_{amm} than mammals. For instance, T_{amm} concentrations in human blood are kept below 40 μM by hepatic urea synthesis, and even small increases (50%) cause neural pathologies. Control T_{amm} concentrations reported for fish blood are up to an order of magnitude higher and more variable (between species and study). Wood attributed much of the variability to different blood collection and analysis techniques and estimated that the "true" control arterial T_{amm} concentrations offishes are <500 μM and typically between 100 and 200 μM. Although fishes are more tolerant of T_{amm} than mammals, they are susceptible to the same toxic effects at higher concentrations and therefore must excrete T_{amm} at the same rate it is synthesized.

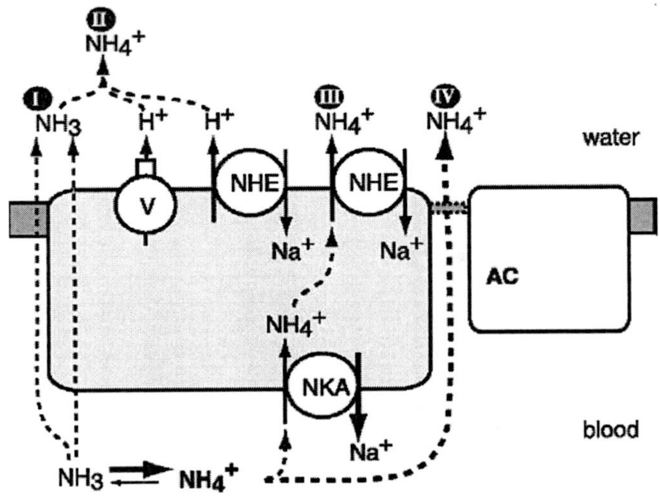

General Ammonia Secretion

FIG. 35. Composite model of ammonia secretion pathways in the gills of fishes (e.g., agnathans, elasmobranchs, and teleosts), numbered in decreasing order of occurrence (Roman numerals). I, *diffusion of NH_3 either through or between epithelial cells down favourable blood-to-water diffusion gradients.* II, *diffusion of NH_3 that relies on proton secretion to "trap" ammonia as NH_4^+.* III, *transport of ammonia through NHE and/or Na^+-K^+-ATPase.* IV, *diffusion of NH_4^+ through the leaky junctions that occur between MRC and AC in seawater teleosts.* Pathway I *can occur for any cell type,* pathways II *and* III *can only occur in acid secretion cells, and*pathway IV *can only occur with salt secretory cells of teleosts.*

Unlike in mammals, where ammonia is synthesized at the site of excretion, i.e., kidneys, <25% of excreted ammonia is synthesized at the gills of fishes. Most of the ammonia excreted by the gills is produced in other tissues and cleared from the blood. During control conditions, the liver appears to synthesize the majority of ammonia in fishes, with skeletal muscle, kidneys, and gills contributing in descending quantitative order. However, skeletal muscle can dominate ammonia synthesis during glycolytic exercise or hypoxia via adenylate deamination. Whole body T_{amm} excretion rates have been measured for many species, including teleosts, elasmobranchs, and agnathans.

Under control conditions, most ammonotelic fishes have excretion rates between 100 and 350 μmol · kg^{-1} · h^{-1}, and ureotelic elasmobranchs have excretion rates below 50 μmol · kg^{-1} · h^{-1}. However, T_{amm} excretion rates over 1,000 μmol · kg^{-1} · h^{-1} have been measured after feeding and NH_4Cl infusion, demonstrating a high capacity for T_{amm} transport across the gill epithelium.

Despite being studied exhaustively over the past 30 years, the exact mechanisms of ammonia transport across the branchial epithelium of fishes remain somewhat controversial, because it is difficult to measure and/or control transepithelial gradients of pH, P_{NH3}, NH_4^+, and electrical potential simultaneously in wholefish or whole tissue experiments. Although at least five potential mechanisms have been hypothesized, with some data supporting each, the majority of current experimental data support three pathways of ammonia excretion from the gills of fishes. These include *1*) NH_3 diffusion, *2*) NH_3 diffusion trapping, *3*) Na^+/NH_4^+ exchange, and *4*) NH_4^+d ffusion. Each mechanism is not exclusive, and their quantitative contributions are determined by conditions such as external water, salinity, pH, and buffer capacity.

NH_3 Diffusion

Experiments on an agnathan, an elasmobranch, and a number of freshwater teleost species suggest that most T_{amm} crosses the branchial epithelium as NH_3, down favourable blood-to-water diffusion gradients (teleost data reviewed extensively by). Three lines of evidence support this conclusion. *1*) Moderate decreases in external water pH, which decreases $[NH_3]$ in the external water via protonation, increased the rate of T_{amm} excretion from freshwater rainbow trout, carp, and goldfish. *2*) Increases in external water pH, which increases $[NH_3]$ in the external water via deprotonation, decreased the rate of T_{amm} from freshwater rainbow trout. *3*) Ammonium injections $[NH_4Cl$ or $(NH_4)_2SO_4]$ caused an initial increase in T_{amm} excretion, in excess of net acid excretion, from Atlantic hagfish, spiny dogfish sharks, rainbow trout, and channel catfish, suggesting that at least some of the infused ammonium dissociated and crossed the gills as NH_3. These ammonium injections also caused a metabolic acidosis, which presumably reflects residual H^+, left in the blood after dissociation of NH_4^+ and excretion of NH_3.

NH_4^+ Diffusion

NH_4^+ diffusion is probably minimal in all fishes except seawater teleosts and lampreys, because of low permeability to cations by the branchial epithelium. Seawater teleosts and lampreys have shallow tight junctions between MRCs, which increase cation permeability for Na^+ secretion and therefore are probably more permeable to NH_4^+ than other fishes. This was suggested by experiments on seawater longhorn sculpin, which did not develop a blood alkalosis when exposed to high environmental T_{amm} concentrations, suggesting that ammonia loading occurred as the acid-base neutral form, NH_4^+, and not the basic form, NH_3. In addition, Goldstein et al. demonstrated that net ammonia excretion rates from seawater longhorn sculpin and toadfish were sensitive to changes in external NH_4^+ concentrations but not external NH_3 concentrations.

Urea

Blood urea concentrations have been measured in many fish species and appear to vary by taxonomic group and by conditions that inhibit passive ammonia excretion via diffusion. Ammonotelic teleosts and agnathans, which include the majority of their respective groups, typically have blood plasma urea concentrations below 2–3 mM, reflecting their low urea synthesis rates and dependence on ammonia excretion to remove nitrogenous waste.

Higher plasma urea concentrations have been measured for teleosts that live in environments that are unfavourable for ammonia excretion. For example, the Lahontan cutthroat trout of Pyramid Lake, a cyprinid of Lake Van, Turkey (*Chalcalburnus tarichi*), and a tilapia of Lake Magadi, Kenya (*Alcolapia grahami*) have plasma urea concentrations of 8.15, 36.18, and 10.52 mM, respectively.

These species all live in waters with a pH between 9.4 and 10.0 that inhibit passive NH_3 excretion via diffusion trapping and appear to elevate urea synthesis to partially (Lahontan cutthroat and cyprinid), or completely (Lake Magadi tilapia), compensate for reduced ammonia excretion. A few other fishes have elevated plasma urea concentrations, either during stress or confinement (gulf toadfish) or air exposure (e.g., lungfishes). By far, the highest blood urea concentrations are found in fishes that are ureosmotic, which include coelacanths and almost all elasmobranchs. The only species of the latter group known to be ammonotelic belong to a single family of stenohaline freshwater stingrays in South America, the Potamotrygonidae.

As with ammonia, the majority of urea appears to be synthesized in the liver offishes, regardless of the enzymatic pathways used to synthesize it. Arginase and uricolytic enzymes synthesize the majority of urea in ammonotelic fishes and appear to be present in most species.

A fully functional ornithine-urea cycle (OUC) appears only in chondrichthyes, coelacanths, and a limited number of teleosts. Because the majority of teleosts are ammonotelic, few studies have investigated mechanisms of urea transport in the gills of teleosts until the past decade. These have largely focused on two species that can be ureotelic and express a fully functional OUC, the gulf toadfish and the Lake Magadi tilapia.

Pulsatile Urea Excretion

Gulf toadfish are normally ammonotelic when they are first acquired from the wild, but soon become facultatively ureotelic (>70% of nitrogen excreted as urea) when crowded or confined in aquaria. Under these conditions, gulf toadfish excrete urea from the gills in discrete pulses that last under 3 h each and occur about one time per day. Remarkably, urea excretion rates rise from near zero (~10 $\mu mol \cdot kg^{-1} \cdot h^{-1}$) during steady-state, nonpulse times to over 300 $\mu mol \cdot kg^{-1} \cdot h^{-1}$ during pulses, and blood concentrations decrease by ~1 mM. This excretion pulse is not due to rapidly increased blood urea concentrations or urea synthesis rates, because neither increases sharply before or during a pulse event. A series of in vivo experiments ruled out two of three hypothetical excretion mechanisms that could explain the observed pulses: *1*) involvement of a normally active back-transport system similar to elasmobranchs and *2*) changes in the general permeability of the gills.

For example, neither Na^+ removal from nor urea analogue addition to the external water increased urea excretion from toadfish during nonpulse steady-state periods, suggesting the absence of a back-transport system as occurs in elasmobranchs. Such treatments would be expected to inhibit a back-transport system and increase urea excretion. A general increase in gill perfusion or permeability was also ruled out by experiments that demonstrated a lack of increased urea excretion after injection of a strong branchial dilator (L-isoprenaline) and by experiments that showed no change in the permeabilities of polyethylene glycol-4000 or water during urea pulses.

Further experiments suggested that the urea permeability could be regulated by periodic translation, insertion, and/or activation of specific urea transporters. For instance, elevating urea concentrations in the external water to three times the blood concentration (30 mM) caused absorption of urea into the blood at each pulse event that was proportional to the relative concentrations in the water and blood, demonstrating that the transport system is reversible. This is a characteristic of rat inner medullary collecting ducts, which express the facilitated urea carrier UT-2.

Walsh et al. now have cloned, sequenced, and functionally expressed a urea transporter from toadfish gills (tUT), which was the first urea transporter sequenced from fish gills. The 1,814-bp cDNA (GenBank accession no.165893) codes for a protein that is over 55% homologous to mammalian, amphibian, and elasmobranch kidney urea transporters and exhibits phloretin-sensitive urea transport when expressed in oocytes. Interestingly, of six tissues assayed with Northern blots including the kidneys, tUT expression was only detected in the gills of toadfish, confirming earlier divided chamber, and pharmacological studies that suggested that gills are the primary site of urea excretion.

Therefore, it was concluded that tUT is the protein responsible for transient urea permeability in toadfish gills. However, no statistical change in tUT expression levels were detected with Northern blots during urea pulses, suggesting that regulation of urea excretion is posttranscriptional.

In support of this hypothesis, Laurent et al. used transmission electron micrographs to demonstrate an increase in vesicle numbers and proximity of vesicles to the apical membranes of PVCs in urea-pulsing toadfish. The authors suggested that tUT might be expressed in these vesicles, which are delivered to the apical membrane during a urea pulse.

Retention Mechanisms in Elasmobranchs

Despite being ureotelic and excreting over 80% of their nitrogenous waste as urea, elasmobranchs require branchial mechanisms to minimise loss of urea that is required to maintain osmotic balance in seawater. For instance, Pärt et al. pointed out that the urea excretion rates of elasmobranchs would be ~10,000 μmol $\cdot$ kg^{-1} $\cdot$ h^{-1} if their gills had urea permeabilities similar to teleosts, which is 40-fold greater than their actual excretion rates (250 μmol $\cdot$ kg^{-1} $\cdot$ h^{-1}). A low urea permeability by elasmobranch gills was first reported for the spiny dogfishby Boylan and later verified by Fines et al. to be 3.2×10^{-8} cm/s, which is over 60-fold less than the permeability of rainbow trout and eel gills. Unfortunately, detailed investigations of the mechanisms responsible for this low permeability have only begun in the past 10 years, and large gaps still exist in the current models. Although the low urea permeability of elasmobranch gills was originally assumed to be a purely structural adaptation, recent data suggest a combination of structural and active transport mechanisms.

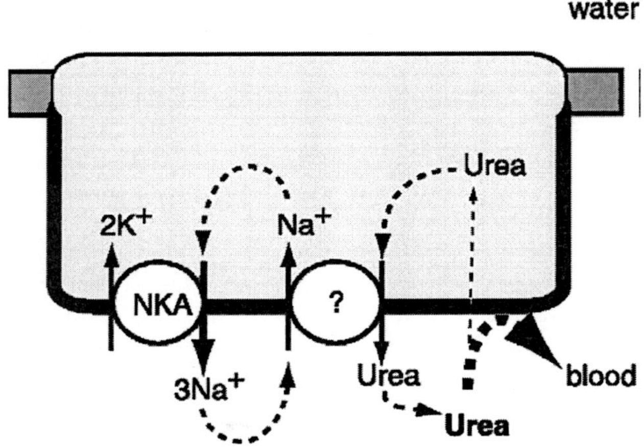

Figure: *Model of urea retention mechanisms proposed for marine elasmobranchs. The basolateral membranes of spiny dogfish shark gills were shown to have a high cholesterol content, which reduces the membrane's permeability to urea (indicated by a thick line for the basolateral membrane). Although this is thought to minimise urea entry into the cells, some still leaks in and is thought to be transported back across the basolateral membrane by an unidentified, Na^+-dependent urea transporter. The Na^+ gradient required for this urea transporter is thought to be maintained by Na^+-K^+-ATPase.*

The structural adaptations may be due to an unusual composition of the plasma membranes in the gill epithelium. For instance, basolateral membrane vesicles (BLMV) from spiny dogfish shark gill tissue were shown to have the highest cholesterol-to-phospholipid ratios ever recorded for a natural biological membrane. This was proposed as a mechanism to minimise urea entry into the epithelial cells across the basolateral membrane, because cholesterol is known to reduce urea permeability by allowing a tighter fit between adjacent phospholipid molecules.

Data from in vivo and isolated perfused head experiments suggested the presence of an active, back-transport system in the gill epithelium spiny dogfish sharks. For instance, urea excretion rates from this species were unaltered by up to 50% changes in urea concentrations of plasma or isolated-head perfusion fluids, suggesting a saturated transport mechanism. In addition, urea excretion rates from spiny dogfish were increased by the competitive urea analogues acetamide and thiourea and by the noncompetitive inhibitor phloretin, suggesting the presence of a carrier-mediated transport system.

Finally, a washout experiment with isolated perfused heads of spiny dogfishsharks demonstrated a 14-fold greater urea transport rate across the basolateral membrane than across the apical membrane. These data support a model of the gill epithelium as an "intermediary compartment" where the urea gradient across the apical membrane is lowered by active transport of intracellular urea across the basolateral membrane.

More recently, a detailed study of BLMV from spiny dogfish shark gills further characterised the putative carrier-mediated transporter. Urea uptake by BLMV was inhibited by the urea analogues *N*-methylurea and nitrophenolthiourea and was inhibited in a dose-dependent manner by phloretin, confirming the presence of carrier-mediated transport. Furthermore, urea uptake was stimulated by ATP, and this stimulation was sensitive to ouabain, suggesting that the urea transporter was secondarily active, dependent on Na^+ and/ or K^+ gradients created by Na^+-K^+-ATPase. Other experiments showed that Na^+ gradients, and not K^+ gradients, stimulated uptake, suggesting that the urea transporter is Na^+ dependent and probably uses the Na^+gradients created by Na^+-K^+-ATPase. Finally, K_m for urea was 10.1 mM, which is much lower than the blood plasma urea concentration. The authors suggested that this may be indicative of an intracellular urea scavenging role for the transporter that actively returns urea to the blood.

4

Neural, Hormonal and Paracrine Control

Given the complexity of the perfusion pathways in the fish gills and the location of specialised transporting cells and paracellular pathways, one might hypothesize that complex control systems have evolved to provide for homeostasis and the potential for rapid and long-term responses to environmental or internal changes. Interestingly, many of these signalling agents have both hemodynamic and ionic transport effects.

Intrinsic Control

Cholinergic Neurons

Stimulation of the "vagosympathetic" nerve trunk to the isolated third gill arch of the Atlantic cod produced an increase in branchial vascular resistance, which was reversed by the addition of atropine, corroborating studies that demonstrated that acetylcholine produced an increase in branchial resistance in the rainbow trout. The early proposition that constriction of postlamellar, efferent vessels underlies this response has been confirmed by in vivo videomicroscopy that visualised constriction of the efferent filamental vessels in the rainbow trout, which was blocked by atropine. The site of constriction is generally assumed to be the prominent sphincter in the proximal portion of the efferent filamental artery that is innervated by vagal cholinergic fibres in a variety of teleost species. Constriction in the postlamellar, efferent filamental artery is accompanied by increased blood flow into the ILV and increased flow through the lamellae, both of which are blocked by atropine. Similar, atropine-sensitive, changes in the perfusion pattern and gill resistance can be elicited by hypoxia, which would increase the functional surface area of the gills, appropriate to compensate for hypoxic conditions. Moreover, diversion of blood into the ILV may provide greater perfusion of MRCs (although no evidence exists suggesting perfusion limitation to transport) and/or plasma skimming to enrich the hematocrit of the efferent filamental blood. Atropine sensitivity suggests a muscarinic receptor in the efferent gill vasculature, but no biochemical or molecular charactérisation of the putative receptor has been published. However, an M_3 type muscarinic receptor has been characterised in the spiny dogfishventral aorta, so it is possible that this receptor mediates the cholinergic effects in the gills also.

To our knowledge only a single study has demonstrated that cholinergic innervation may have a direct effect on ionic transport across the gill epithelium. May and Degnan found that acetylcholine inhibited the I_{sc} across the isolated killifish opercular epithelium.

Adrenergic Neurons and Chromaffin Tissue

Epinephrine and norepinephrine reach gill tissues via both neurotransmission from sympathetic innervation (apparently only in teleosts), and from stimulation of chromaffin cells in the teleost and elasmobranch interrenal gland, which is associated with the head kidney. Intravenous infusion of catecholamines into a variety of teleost species is associated with a biphasic response in gill resistance: an initial, á-adrenergic-mediated increase, followed by a longer-lasting, β-adrenergic-mediated fall in resistance. Stimulation of the sympathetic chain anterior to the celiac ganglion in the cod produced the same á-mediated vs. β-mediated responses.

The -adrenergic-mediated fall in gill resistance was associated with an increased oxygen uptake and was presumed to be the result of lamellar recruitment. Accordingly, hypoxia elicits an increase in plasma catecholamine levels, with the attending changes in gill perfusion, as part of a complex response to maximise oxygen delivery to the tissues. Measurement of branchial venous outflow in the cod in vivo has demonstrated that the á-mediated increase in resistance is secondary to constriction of the arteriovenous anastomoses between the efferent filamental artery and the ILV, where nerve terminals have been observed. The β-mediated fall in branchial resistance is generally considered to be the result of dilation of afferent lamellar arterioles, which produces lamellar recruitment. It is noteworthy that catecholamines also produce preferential arterio-arterial flow in the hagfish gill pouch, suggesting an ancient origin for at least the sympathetic, branchial control mechanisms.

In addition to potential effects of catecholamines on gill permeability and transport produced by changes in perfusion patterns, it is clear that these vasoactive mediators can have direct effects on branchial ionic transport. The initial characterisation of the killifish operculum as a model for teleost gill salt extrusion demonstrated that epinephrine inhibited the I_{sc} produced by the tissue. Subsequent studies demonstrated that Cl^- fluxes across the tissue could be inhibited by á-adrenergic activation and stimulated by β-adrenergic activation and that inhibition versus stimulation of the I_{sc} was via $á_2$-adrenergic and and $β_1$-adrenergic receptors, respectively. Moreover, cAMP is the second messenger for the β-stimulation. A more recent study has demonstrated that stimulation of the trigeminal nerve associated with the opercular epithelium produced a á-mediated inhibition of the I_{sc} with an associated increase in intracellular inositol trisphosphate (IP_3) levels. Adrenergic neurons may also play a role in modulating Ca^{2+} balance, because one study demonstrated that epinephrine injection or stimulation of adrenergic branchial nerves inhibited $^{45}Ca^{2+}$ uptake into the gill tissue of the rainbow trout.

Serotonergic Neurons and Neuroepithelial Cells

The fact that the constrictory response to nerve stimulation in the cod could not be totally abolished in some preparations by pretreatment with either 10 μM atropine or phentolamine (á-adrenergic antagonist) suggested that some nonadrenergic, noncholinergic (NANC) fibres may exist in the gills. Serotonergic fibres now have been described in the gills of a variety of teleost species, and serotonin increased branchial resistance in a number of species in vivo and in vitro. Video observation of blood flow through the rainbow trout gills has shown that serotonin redistributes blood flow to the more proximal parts of the filament, by constriction of distal portions of the efferent filamental artery and possibly the afferent filamental artery.

The resulting reduction of lamellar perfusion presumably accounts for the reduction in gas exchange when serotonin is infused into two species of teleosts. In the cod, serotonin also dilates the arteriovenous anastomoses, shunting more blood into the ILV. Interestingly, serotonin apparently stimulates the release of catecholamines in at least the rainbow trout. Sensitivity to methysergide suggests that the serotonin effect is mediated by 5-HT_2 receptors in at least three species of teleosts. Serotonin is also produced in gill neuroepithelial cells, which are on efferent side of gill filaments and assumed to be O_2 sensors, but the relative roles of NEC release versus neuronal release are unclear.

To our knowledge the effects of serotonin on fish gill transport are unstudied, but the nonmetabolised, serotonin agonist 8-hydroxy-2-(di-*n*-propylamino)tetraline (8-OH-DPAT) stimulates Na^+-K^+-ATPase expression and enzymatic activity in the opossum proximal tubule (61). Therefore, similar studies in fish gills would be interesting.

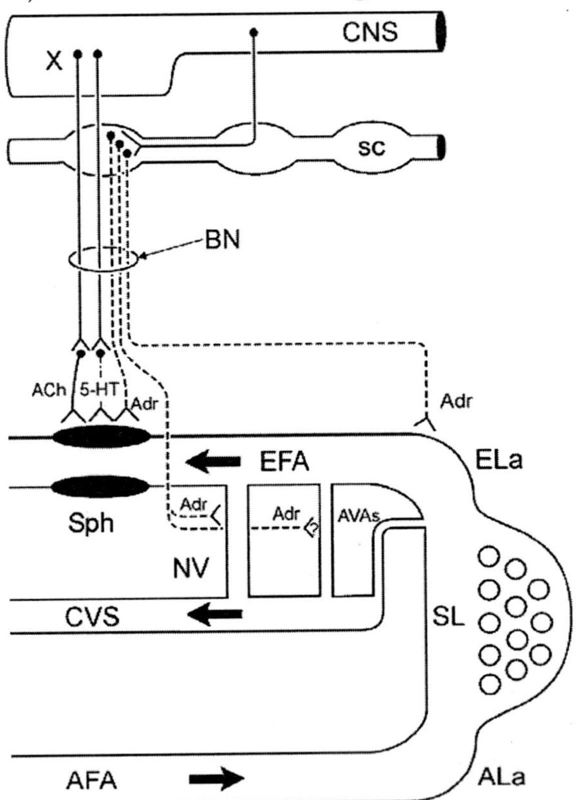

Figure: *Working model for the autonomic control of teleost gill vasculature. The sphincter (Sph) at the base of the efferent filamental arteries (EFA) is the site of cholinergic (ACh), adrenergic (Adr), and serotonergic (5-HT) innervation. ACh and 5-HT innervation provide constriction, whereas Adr innervation may cause dilation, via β-adrenoceptors. Other potential sites for perfusion control are the nutritive vasculature (NV) and arteriovenous anastomoses (AVAs). Adr innervation of NV, and possibly AVAs, produces vasoconstriction via α-adrenoceptors. SLA, secondary lamella (=lamella); BN, branchial nerve; CNS, central nervous system; sc, sympathetic chain, X, vagus nerve.*

Nitrergic Neurons and Neuroepithelial Cells

Branchial NANC innervation may also include nitrergic fibres, because NADPH-diaphorase reactivity was seen in 55–85% of branchial nerves in the cod, as well as in the putative sympathetic trunk (which supplies the gill arches) in stage 13 (4.5 days of development) of the spotted tilapia (*Tilapia mariae*). NADPH-diaphorase-reactive nerves also have been described in the vasculature in the gills of the Indian catfish. It has been suggested that NADPH-diaphorase histochemistry may not be as specific for nitric oxide synthase (NOS) as formerly assumed, but immunoreactive nitrergic fibres have now been described associated with the efferent filamental arteries in the Indian catfish. In addition, neuronal NOS (nNOS) immunoreactivity is seen in neuroepithelial cells adjacent to the efferent filamental artery along the length of the filament in the Indian catfish, in the interlamellar region in both the killifish and longhorn sculpin (Hyndman and Evans, unpublished data), and in cells distinct from but adjacent to Na^+-K^+-ATPase-containing MRCs in the operculum of the killifish. Partial clones for nNOS from the gills of the spiny dogfish (GenBank accession no. 232227) and killifish (GenBank accession no. AY533030) now have been reported.

There is clear evidence that either a NO donor (sodium nitroprusside, SNP) or NO itself is vasodilatory in systemic vessels (and the ventral aorta) in the rainbow trout and eel, and infusion of NOS inhibitors N^G-nitro-L-arginine methyl ester (L-NAME) or N^G-nitro-L-arginine (L-NNA) constricted the rainbow trout coronary system. On the other hand, SNP constricted the ventral aorta, anterior mesenteric artery, and posterior intestinal vein of the dogfish shark and the ventral aorta of the hagfish and produced constriction followed by dilation in the ventral aorta of the sea lamprey. In contrast to the finding that SNP dilated the eel ventral aorta, another study has found that SNP and SIN-1 (another NO donor) contracted the perfused gill preparation of the same species. The effect could be mimicked by the stable cGMP analogue 8-bromo-cGMP and inhibited by the addition of the soluble guanylyl cyclase (sGC) inhibitor 1H-[1,2,4]oxadiazole[4,3-a]quinoxalin-1-one (ODQ), demonstrating that intracellular cGMP was involved. To our knowledge, there are no published data suggesting that NO affects gill vasculature directly, but it likely does, given the extensive expression of nNOS in the gills. A more specific description of NO effects on gill vessels awaits the kind of video-microscopic studies that have been described for other effectors. Despite obvious vasoactivity of NO in fish vessels, it appears that a prostanoid, not

NO, is the endothelium-derived relaxing factor in at least the rainbow trout and spiny dogfish, so it appears that branchial NO is likely produced by the NOS-containing neurons and neuroepithelial cells that have been described above.

Recent evidence suggests that NO may have a role in modulating salt extrusion by the gill epithelium. For example, if L-NAME was applied to the killifishopercular epithelium, the I_{sc} increased by 15%, suggesting tonic, inhibitory control by NO. In addition, inhibition of NOS by L-NAME reduced the inhibitory effect of subsequent addition of sarofotoxin S6c [endothelin (ET) agonist], suggesting that a component of the ET-mediated inhibition is secondary to stimulation of NO production.

Adenosine

The paracrine adenosine also may affect gill blood flow. Adenosine increased branchial vascular resistance in rainbow trout, and this was associated with a vasoconstriction of the efferent filamental artery and shunting of blood to the ILV and more proximal parts of the filament, which was mediated by an A_1-type receptor. Activation of A_1 receptors also increased gill resistance (and decreased oxygen uptake) in the Antarctic borch, but adenosine did not affect the resistance in perfused hagfish gills, although it did reduce systemic resistance. Adenosine agonist assays demonstrated the presence of both A_1 (constrictory) and A_2 (dilatory) receptors in the ventral aorta of the dogfish shark so it is possible that both receptors may be in the gill vasculature. We are unaware of any published studies on the putative effects of adenosine agonists on gill epithelial transport, as has been described for ionic transport in the mammalian renal tubules.

ET

ET is as potent in constricting fish vasculature as it is in mammals, and the effects are both systemic and branchial. An early study demonstrated that rat ET-1 increased gill resistance in the rainbow trout, and rainbow trout ET (isolated from rainbow trout kidney extracts) potently (and concentration dependently) constricted a variety of rainbow trout vascular rings, including the efferent branchial artery. Interestingly, the trout vascular preparations were more sensitive to rat ET-1 than the homologous peptide, but the maximum tensions produced by both peptides were equivalent. Heterologous ET-1 also produced an increase in branchial resistance in the cod and constricted systemic vessels in two agnathans, the spiny dogfish and the eel. Agonist sensitivity suggests that the gill ET receptor in the spiny dogfish is an ET_B type. Lodhi et al. suggested that the ET receptor in the gills of the rainbow trout was not a classical ET_A type, because of its insensitivity to BQ-123, but they did not test for an ET_B-like receptor. They localised the ET receptor to the lamellae by ^{125}I-labelled ET-1 autoradiography, which is consistent with the subsequent, video-microscopic demonstration that ET-1 constricted lamellar pillar cells in both the rainbow trout and cod, thereby shunting blood to the outer margins of the lamellae. Our recent video-microscopic studies with the longhorn sculpin have demonstrated a nearly complete shutdown of flow to the distal filament and lamellae, consistent with pillar cell contraction and possibly constriction of the afferent filamental artery or prelamellar arterioles. Moreover, we have immunolocalised ET_B-like receptors to the afferent and efferent filamental arteries in the gills of the killifish and longhorn sculpin (as well as the postlamellar, arteriovenous anastamoses), but not to the pillar cells (K. A. Hyndman and D. H. Evans, unpublished data). Thus species differences may exist in the specific distribution of ET receptors in the fish gills. The site of ET synthesis appears to be vascular endothelial and neuroepithelial cells in the filamental epithelium in a variety offish species including teleosts and elasmobranchs. In Zaccone's studies, big ET (ET precursor) immunoreactivity colocalised with serotonin and neuropoeptide immunoreactivity in some neuroepithelial cells. Big ET has been localised to Na^+-K^+-ATPase-containing MRC in the gill epithelium of the Atlantic stingray.

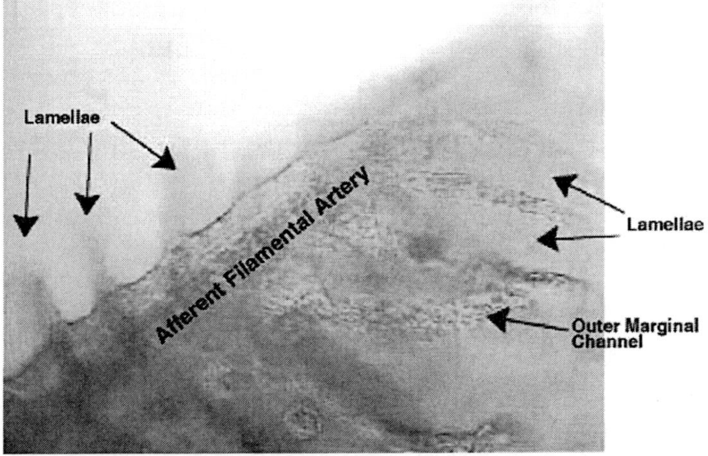

Figure: *Video of the effect of endothelin on the flow of blood through the afferent filamental artery and a series of afferent lamellar arterioles leading into lamellae of a single filament in the gill of the longhorn sculpin* Myoxocephalus octodecimspinosus.

Endothelin has direct effects on mammalian renal transport, and recent data suggest that epithelial effects of ET can be demonstrated in fishalso. Specifically, ET-1 inhibited the I_{sc} produced by the killifish opercular epithelium in a concentration-dependent manner, and the effect is mimicked by the ET_B-specific agonist sarafotoxin S6c, suggesting a role for ET_B-like receptors.

Prostanoids

Branchial production of prostanoids may play a role in gill perfusion, because a cyclooxygenase (COX) inhibitor, indomethacin, attenuated the VIP-induced dilation of perfused rainbow trout gills. However, the data on the vasoactivity of prostanoids in fish vessels are conflicting. For example, Piomelli et al. demonstrated a prostacyclin (PGI_2)-induced decline in gill vascular resistance in perfused heads of two elasmobranchs, but an increase in resistance in the same preparation in four teleosts, as well as perfused arches of two other teleosts and an elasmobranch.

In all cases, adrenergic receptor inhibitors had no effect, suggesting that the effects were direct and not mediated by secondary release of epinephrine. A subsequent study showed that PGE_2 increased gill resistance in eels in vivo, and this has been corroborated by a video study that showed that infusion of $PGF_{2\alpha}$, but not PGE_2, decreased the diametre of the afferent filamental arteries in the cod. In all, it appears that the prostanoids are constrictory in the gills and may have prelamellar effects on the afferent filamental artery. Gill tissue from both the rainbow trout and European eel produces prostanoids, including PGE_1, PGE_2, $PGF_{1\alpha}$, $PGF_{2\alpha}$, and PGD_2, and in the eel (but not rainbow trout) the levels of each fell when the fishwere acclimated to seawater. Using a heterologous antibody, COX immunoreactivity has been localised to the ILV of the Atlantic stingray gills and in MRCs in the killifish gills (Piermarini and Evans, unpublished data), and we have recently cloned a COX-2-like fragment from the gill of the killifish.

Prostanoids (specifically PGE_2) also may play an important role in modulating NaCl extrusion by the marine teleost gills. Two early studies demonstrated that PGE_2 inhibited the I_{sc} across the killifish opercular epithelium and a recent study found that PGE_2 inhibited the I_{sc} produced by a cultured epithelium of PVCs from the sea bass (*Dicentrarchus labrax*). However, it should be noted that this cultured epithelium is presumably lacking MRCs and produces a very small I_{sc} compared with the killifish opercular preparation.

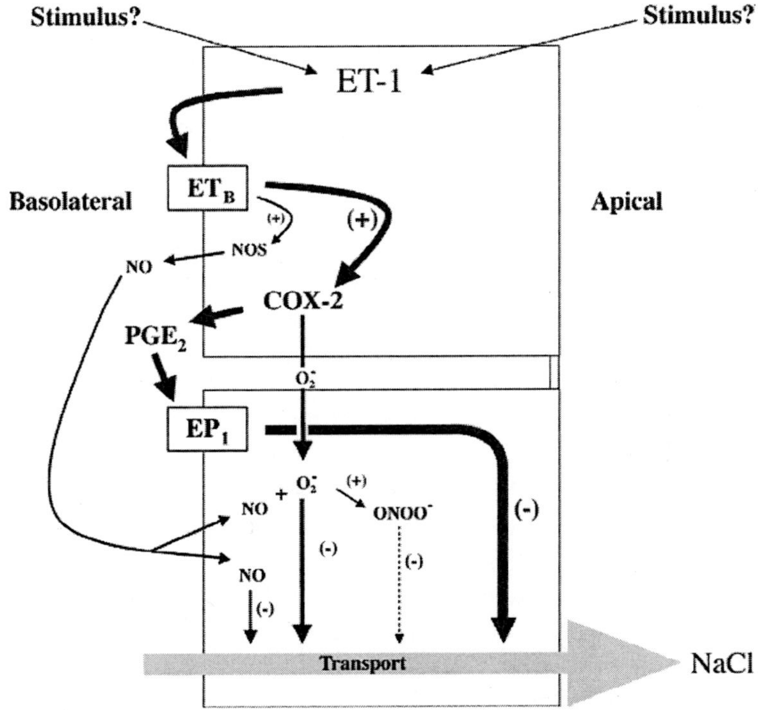

Figure: *Working hypothesis for the putative pathways of endothelin (ET) inhibited NaCl transport across the fish gill. Two cells are diagrammed, but the system may be expressed within a single cell. The width of the arrows is proportional to the presumed importance of the specific pathway in the axis. ET-1, endothelin; ET_B, endothelin B receptor; NOS, nitric oxide synthase; COX-2, cyclooxygenase-2; NO, nitric oxide; PGE_2, prostaglandin E; EP_1, PGE_2 receptor; O_2^-, superoxide ion; $ONOO^-$, peroxynitriteion.*

As mentioned above, the ET-1 (actually the ET_B agonist sarafotoxin S6c) mediated reduction of the I_{sc} across the killifish opercular epithelium can be inhibited partially by preincubation of the tissue with either L-NAME (17%) or TEMPOL (34%), suggesting that both NO and O_2^- may be secondary mediators of the ET response. Preincubation of the tissue with indomethacin inhibited the sarafotoxin S6c effect by 90%, suggesting that COX-produced prostanoids are the major effectors of this axis, just as they appear to be the dominant endothelium-derived relaxing factor (EDRF) in fishes. Moreover, PGE_2 produced a concentration-dependent inhibition of the I_{sc}. Other prostanoids were not inhibitory. The COX-2 specific inhibitor NS-398 was much more effective than the COX-1 inhibitor SC560, and an $EP_{1/3}$-receptor specific agonist, sulprostone, produced a concentration-dependent inhibition of the I_{sc}, but butaprost (an EP_2-specific agonist) produced a concentration-dependent stimulation. A working model for this new signalling axis in the fish gill is diagrammed below.

Extrinsic Control

Prolactin

We are unaware of any studies on the hemodynamic effects of prolactin in fishes, but it has been clear for 45 years that the adenohypophysial peptide plays a major role in gill function in freshwater osmoregulation. In a classic paper, Pickford and Phillips demonstrated that hypophysectomised killifishcould survive in fresh water only if they were treated with prolactin, and subsequent isotopic studies showed that this survival was associated with a reduction in ionic loss, rather than stimulation of ionic uptake.

More recently, it has been shown that prolactin can also restore osmoregulatory ability to the hypophysectomised channel catfish in freshwater. Prolactin injections reduced gill Na^+-K^+-activated ATPase activity in the killifish, the thick-lipped grey mullet (*Chelon labrosus*), gilthead sea bream, and rainbow trout. However, prolactin had no effect on either expression of Na^+-K^+-ATPase (by Northern blot) or enzymatic activity in the gill epithelia of the brown trout and tilapia, as well as the opercular epithelium in tilapia.

In the latter study, prolactin did not affect MRC number, but it reduced the size of individual cells, and this was corroborated by a subsequent study in a related species (*O. niloticus*) that found that prolactin injection stimulated the formation of small MRCs, similar to the "beta cells" of freshwater acclimated teleosts. A similar cellular remodelling was produced by prolactin injection into the silver sea bream (*Sparus sarba*). These studies corroborate earlier work that found that prolactin injections into tilapia were associated with a reduction of both I_{sc} and conductance across the in vitro opercular epithelium, suggesting epithelial remodelling in addition to direct effects on ionic transport.

As one might expect, plasma levels of prolactin are inversely related to salinity. A recent study has demonstrated the importance of prolactin in freshwater ion regulation directly, by showing that a cultured gill epithelium (that contains both PVCs and MRCs) can only transport Na^+ and Cl^- inward when prolactin and cortisol are in the culture medium. Interestingly, Na^+-K^+-activated ATPase activity was unaffected by these treatments.

Prolactin receptors have been cloned in several fish species, including two species of tilapia and rainbow trout, goldfish, and gilthead sea bream, and expression was high in the gill in all cases. Using either in situ hybridisation or immunohistochemistry, the receptors could be localised to the MRCs in most of these species. Interestingly, in the tilapias and rainbow trout, transfer to hyperosmotic salinities was associated with equivalent or increased expression of the receptor, suggesting that prolactin might have some role in acclimation to salinities other than fresh water.

Growth Hormone and IGF

As might be expected, fish growth hormone (GH) has morphogenic effects (independent of GH-enhanced size), and the major effect appears to be stimulation of osmoregulation in seawater. Injection of GH promotes salinity tolerance, associated with increased Na^+-K^+-activated ATPase activity and MRC size and density, in salmonids as well as two species of tilapia, the killifish, and silver sea bream. GH injections were associated with increased expression of Na^+-K^+-ATPase in MRC of the brown trout and expression of NKCC1 in the MRCs in the gills of the Atlantic salmon. GH receptors have been localised to gills in the coho salmon by radioreceptor assay, and two splice variants of a recent clone from the black sea bream (*Acanthopagrus schlegeli*) can be localised to a variety of tissues, including the gills.

In the killifish, Atlantic salmon, and brown trout, some of the effects of GH can be mimicked by injection of IGF. For example, IGF-I injection stimulated MRC development and Na^+-K^+-ATPase expression in the MRC in the gills of the brown trout. Transfer to seawater is associated with an increase in plasma concentrations of GH in salmonids, as well as increased message for IGF-I in the gills of coho salmon. The gene for IGF-I and its product can be localised to MRC in the tilapia gills, and an IGF-I receptor has been cloned and localised to the gills in two sculpins. It is generally considered that IGF-I is the direct mediator of increased Na^+-K^+-ATPase activity in the MRCs but that GH is needed for the response.

The fact that cortisol and GH together stimulate salinity tolerance and Na^+-K^+-activated ATPase activity better than either one alone in various salmonids and the killifish (and NKCC1 expression in the Atlantic salmon) suggests that cortisol and GH interact to promote seawater osmoregulation. There also appears to be some synergism between cortisol and IGF-I injections. One site of GH/cortisol interaction may be cortisol receptors because GH treatment was associated with an increase in the number of cortisol receptors in the gills of the coho and Atlantic salmon. The synergism may be cytogenic, however, because GH increased general mitotic activity in the gills of the rainbow trout, but cortisol specifically stimulated the number of MRC.

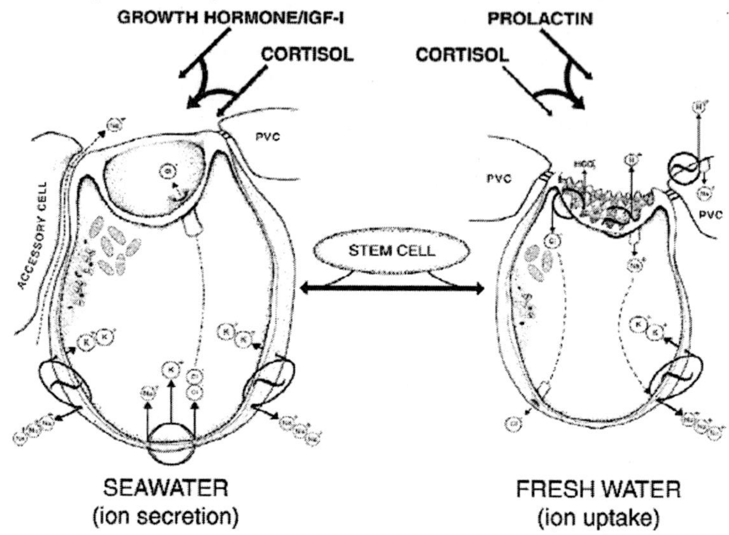

Figure: *Working model for the effects of prolactin, cortisol, growth hormone, and insulin-like growth factor I on the morphology and transport mechanisms of gill mitochondrion-rich cells in the gill epithelium of teleost fish. In seawater, MRCs are generally larger and contain a deep apical crypt; in fresh water, the apical surface is broad and contains numerous microvilli. Growth hormone and cortisol can individually promote the differentiation of the seawater MRC, and also interact to control epithelial transport capacity. Prolactin inhibits the formation of seawater MRCs and promotes the development of freshwater MRCs. Cortisol also promotes acclimation to fresh water by maintaining ion transporters and MRCs, and by interacting to some degree with prolactin.*

Angiotensin

Despite earlier uncertainty, it is now clear that a renin-angiotensin system (RAS) is expressed in fishes, including elasmombranchs and agnathans, although fish produce little or no aldosterone. ANG II constricted perfused gills in both the trout and eel, but local, tonic production of ANG II appears to be relatively unimportant in gill hemodynamics, because inhibition of ANG II production in the trout in vivo by the angiotensin converting enzyme (ACE) inhibitor lisinopril did not affect gill resistance, although systemic resistance was reduced by nearly 50%.

Fish ANG II has three amino acid substitutions in the basic octapeptide, compared with mammals. There generally is a direct correlation between plasma ANG II (and renin) levels and salinity in fishes, including the river lamprey, two eel species, and two salmonids. Enzymatic analyses have found ACE activity in the gills of the trout and river lamprey as well as in the corpuscles of Stannius in the rainbow trout and American eel. Fish corpuscles of Stannius are located within the renal tissue of teleosts and are the site of the production of a unique hormone, stanniocalcin, important in Ca^2 regulation by the gills. Within the rainbow trout gill, ACE can be localised to pillar cells nearest the inner lamellar margin and, to a lesser extent, the lumen of the afferent filamental artery and afferent lamellar arterioles.

ANG II receptors have been localised to gill tissue in the banded houdshark (*Triakis scyllia*) by radioligand binding, and antibodies raised against the mammalian AT_1 receptor localised the receptor to MRCs and PVCs in the gill epithelium in the eel and icefish (*Chionodraco hamatus*). A partial sequence for the eel ANG II receptor has been reported (Tran Van Chuoi et al., GenBank accession no. AJ005132), but no expression data are available. In the eel, acclimation to seawater was associated with a threefold increase in immunoreactive AT_1 in the gills. Infusion of ANG II into the European eel was associated with a concentration-dependent increase in Na^+-K^+-activated ATPase activity in the gills of freshwater acclimated fish, but a biphasic (stimulation inhibition) response with the same concentrations in the seawater acclimated fish. This may appear to be counter to the finding that ANG II inhibited the transepithelial potential across perfused flounder gills, but it is generally accepted that the transepithelial potential in vivo is the result of differential gill ionic permeabilities, not active transport. Unfortunately, no data have been published on the effect of ANG II on the I_{sc} across the killifish opercular epithelium.

Natriuretic Peptides

Various components of the natriuretic peptide (NP) system are expressed in all fish groups that have been examined to date (including the elasmobranchs and agnatha), and NPs play important roles in cardiovascular, renal, and gill physiology. Amino acid sequences have been published for atrial NP (ANP) in the Japanese eel (*Anguilla japonica*) and rainbow trout, and for C-type NP (CNP) in four species of elasmobranchs and three species of teleosts. Four distinct CNPs have been cloned recently from the medaka (*Oryzias latipes*) and fugu. Brain NP (BNP) has not been isolated from fish, but a unique NP, termed VNP, has been isolated from ventricular tissue from the Japanese eel and two species of salmonids. Teleosts produce both ANP and CNP, but elasmobranchs apparently express only CNP, and a recent study suggests that the agnathan NP is unique. Four NP receptors have been cloned from the Japanese eel (termed NPR-A through D), with mammalian-like structures and ligand specificities in the first three. NPR-D has only been described in fishes and is closely related to NPR-C, sharing a short cytoplasmic carboxy-terminal tail that does not include the guanylyl cyclase catalytic domain. NPR-B has been cloned from the rectal gland of the spiny dogfish shark and from the medaka, and partial sequences for an NPR-A/B and NPR-C/D have been reported from the New Zealand hagfish (*Eptatretus cirrhatus*) and pouched lamprey, respectively. A partial sequence for a putative NPR-C/D also has been reported for the spiny dogfish. Radioligand binding protocols have localised NPR-A/B and NPR-C / D receptors in the gills of gulf toadfish, Atlantic hagfish, pouched lamprey, and New Zealand hagfish. It is generally assumed that the CNP/NPR-B couple is the ancestral vertebrate condition.

It has been pointed out that the use of heterologous antibodies for radioimmunoassay determination of plasma NP concentrations often has produced conflicting results, and recent results, using homologous antibodies for Japanese eel peptides, have shown that salinity changes in this species are associated with differential secretion of specific NPs. For instance, the hypernatremia associated with acclimation to seawater is associated with an increase in ANP secretion (and a secondary stimulation of cortisol secretion); acclimation of eels to freshwater is associated with an upregulation of both CNP expression and its specific receptor, NPR-B. NPs are potent dilators of vascular smooth muscle in fishes, including teleosts, elasmobranchs, and agnathans, and ANP decreased gill resistance in the rainbow trout in vivo, corroborating an earlier study that demonstrated that ANP decreased arterio-arterial resistance in the epinephrine-contracted, perfused gills from that species. Unfortunately, no in vivo videomicroscopy studies have been done to visualise the effects of NPs on gill blood flow directly.

Since an early study found that a truncated ANP stimulated the I_{sc} across the killifish opercular epithelium it has been accepted generally that one of the prime roles of NPs in seawater teleosts is the stimulation of gill Na^+ excretion by the branchial epithelium.

However, a recent study was unable to elicit any effect on the I_{sc} across this epithelium by either eel ANP or porcine CNP (0.1 nM to 0.2 μM), so it is not clear that NPs stimulate salt extrusion by the marine teleost gills, despite the fact that ANP apparently may increase the Na^+-K^+-activated ATPase activity in isolated gill cells from the eel (Flik and Takei).

Thyroid Hormones

No data have been published suggesting that thyroid hormones affect gill perfusion, but they may play an indirect, stimulatory role in gill salt extrusion. For instance, thyroxine (T_4) or triiodothyronine (T_3) treatment stimulated Na^+-K^+-activated ATPase activity and/or MRC number in the Atlantic salmon, rainbow trout, climbing perch, and tilapia , as well as immunoreactive Na^+-K^+-ATPase in MRC in the gills of the summer flounder (*Paralichthys dentatus*). However, it appears that these effects may be mediated via the cortisol and/or GH-IGF axes, although at least one study found that the effects on Na^+-K^+-activated ATPase activity of GH and T_3 were additive. For instance, T_4 treatment alone did not stimulate Na^+-K^+-activated ATPase activity in the gills of tilapia, but it potentiated the stimulation produced by cortisol, as it did the effect of GH on enzyme activity in the amago salmon (*Oncorhynchus rhodurus*). Moreover, T_3 treatment increased the number of cortisol receptors in the gills of the Atlantic salmon, which was further potentiated with GH addition, and it increased the number of cortisol receptors (and the cortisol stimulation of Na^+-K^+-activated ATPase activity) in rainbow trout gills.

Glucagon

As is true for thyroid hormones, pancreatic glucagon has not been shown to affect gill perfusion, but this peptide may play a role in gill transport. In an early study of the tilapia opercular epithelium, Foskett et al. found that glucagon could stimulate the I_{sc} across the epithelium in a concentration-dependent manner, with a minimum effective dose of 1 nM and a maximum stimulation of 72% at 10 μM. Moreover, glucagon stimulated adenylate cyclase in the rainbow trout gills, as it does in the process of stimulating NHE3 in the mammalian kidney, a transporter of relevance to fish gills.

Calcitonin and Calcitonin Gene-related Peptide

Calcitonin (CT) is produced in the ultimobranchial glands of fishes and tetrapods, except for mammals, where it is synthesized in the parafollicular "C" cells in the thyroid gland. Agnathans lack the gland,

and they were thought to lack CT until it was recently immunolocated in the plasma of freshwater (but not seawater)-acclimated, Japanese lamprey (*Lampetra japonica*) and also in the inshore hagfish (*Eptatretus burgeri*), where concentration was correlated with plasma Ca^{2+} levels. Calcitonin immunoreactivity has been measured in a number of species of teleosts and elasmobranchs and the peptide sequence has been determined for a number of species, including fugu.

The CT gene also has been sequenced from many species, and the gene and its transcript can be localised to the gills of the Atlantic salmon, so the ultimobranchial gland may not be the only site of CT production. Moreover, a CT-receptor like gene has been cloned from the gill cDNA library of the Japanese flounder (*Paralichthys olivaceus*).

Calcitonin is hypocalcemic in tetrapods, and current evidence suggests that this also is the case in fishes, although hypercalcemic effects are sometimes described.

Infusion of Ca^{2+} into the Japanese eel was associated with an increase in plasma CT immunoreactivity, but transfer of this species from freshwater to seawater did not elicit a change in plasma CT levels, despite increased plasma Ca^{2+} concentrations. However, transfer of rainbow trout to seawater was associated with significant increases in plasma CT concentrations. Several studies have demonstrated that CT inhibits Ca^{2+} uptake by the gills, consistent with a hypocalcemic role.

Calcitonin gene-related peptide (CGRP), a splice variant of the CT mRNA, is also found in fishes and may play a role in vascular dynamics, osmoregulation, Ca^{2+} balance, and gas exchange. CGRP-containing fibres have been described in the brain and/or gut of all the major fish taxa. The peptide also has been identified in the plasma of the pink salmon (*Oncorhynchus gorbuscha*) and in gills of the rainbow trout by radioimmunoassay, where it binds specifically and stimulates adenyl cyclase by a GTP-binding process.

The gene for a CGRP receptor has been cloned from the gills of the Japanese flounder. Acclimation of the rainbow trout to seawater is associated with an increase in plasma levels of CGRP as well as specific binding to gill membrane receptors. Interestingly, this study also found increased gill CA levels in seawater, and this group previously had demonstrated that CGRP can stimulate production of this enzyme in trout gill extracts. Finally, it is possible that CGRP could be vasodilatory in the fishgills, as it is in small arteries from the rainbow trout gut.

Parathyroid Hormone-related Protein

Fishes lack an encapsulated parathyroid gland, but parathyroid hormone (PTH)-like immunoreactivity has been reported in the pituitaries and plasma of several teleosts. However, the recently described PTH-related protein (PTHrP) is also a hypercalcemic factor in humans, shares some common regions at the amino terminus, and binds to a common PTH/PTHrP receptor. PTHrP-like immunoreactivity has now been localised in the pituitary, plasma, or other tissues in a variety of fishes, including teleosts, elasmobranchs, and agnathans, and immunoreactivity was localised to the gill epithelium in some of these studies, with expression confirmed by molecular techniques. The gene for PTHrP was initially cloned from fugu, but a sequence now has been reported from the gilthead sea bream, and in situ hybridisation has localised expression to, among other sites, MRCs in the gill epithelium. A radioimmunoassay also has been developed and validated for the peptide transcribed from this sequence, and immunoreactivity was localised to the gills, as well as the pituitary, esophagus, kidney, and intestine. Two relevant receptors have now been cloned from the zebrafish, one more sensitive to fugu PTHrP than human PTH and another sensitive to human PTH and not human PTHrP.

In an unpublished study, it was shown that PTHrP-like immunoreactivity in the plasma of the European flounder was inversely correlated with salinity (highest in gill tissue from freshwater acclimated individuals), suggesting that the peptide may play some role in Ca^{2+} regulation, which differs with salinity. Exposure of larval gilthead sea bream to an amino-terminal peptide from the fugu sequence increased $^{45}Ca^{2+}$ influx in both 100 and 33% seawater-acclimated fish and reduced Ca^{2+} efflux from the seawater-acclimated fish, both in a concentration-dependent manner. To our knowledge, these are the only published data demonstrating that PTHrP is a functional hypercalcemic factor in fishes.

Other Hypercalcemic Factors

In artificial, calcium-deficient seawater, hypophysectomised killifish displayed reduced plasma Ca^{2+} levels, without any discernable changes in other plasma electrolytes, and the hypocalcemia was not seen in seawater with normal Ca^{2+}levels or after replacement therapy with either pituitary homogenates or heterologous prolactin. Ovine prolactin injections were associated with hypercalcemia in the intact European eel, stimulation of Ca^{2+} uptake from the fresh water, and increased branchial Ca^{2+}-activated ATPase activity. Similar results have been published for tilapia and carp. As one might expect for a putative hypercalcemic factor, pituitary prolactin content is often inversely related to external Ca^{2+} concentrations.

Somatolactin (SL) is a novel pituitary protein that was initially isolated from the pars intermedia of the cod and has been described in the pituitary of a variety of fish species. It is structurally related to both prolactin and growth hormone and was initially described only in teleosts, but has more recently been found in the African lungfish pituitary, suggesting that it may be described in tetrapods in the future. The fact that low environmental Ca^{2+} levels are associated with activation of SL cells, and increased mRNA levels, in the pituitary of the rainbow trout (as well as plasma SL levels) suggests that SLs is a hypercalcemic factor. Somatolactin also may function in reproduction, acid-base regulation, and stress responses in fishes.

Cortisol also may be a hypercalcemic factor in fishes, because plasma concentrations of cortisol increased when the rainbow trout was acclimated to freshwater with a reduced Ca^{2+} concentration, and 8 days of this treatment resulted in a stimulation of branchial Ca^{2+} transport and Ca^{2+}-activated ATPase activity. In freshwater with normal Ca^{2+} concentrations, treatment with cortisol was associated with a stimulation of Ca^{2+} uptake, branchial transport capacity, and hypercalcemia. Although these effects may be related to the cytogenic effects of cortisol, it has been shown recently that cortisol treatment is rapidly (30 min.) followed by a stimulation of gill Ca^{2+}-activated ATPase activity, suggesting that the cortisol effect is not via changes in transcription.

Stress Response in Fish

The stress response in teleost fish shows many similarities to that of the terrestrial vertebrates. These concern the principal messengers of the brain-sympathetic-chromaffin cell axis (equivalent of the brain-sympathetic-adrenal medulla axis) and the brain-pituitary-interrenal axis (equivalent of the brain-pituitary-adrenal axis), as well as their functions, involving stimulation of oxygen uptake and transfer, mobilisation of energy substrates, reallocation of energy away from growth and reproduction, and mainly suppressive effects on immune functions. There is also growing evidence for intensive interaction between the neuroendocrine system and the immune system in fish. Conspicuous differences, however, are present, and these are primarily related to the aquatic environment of fishes. Toxic stressors are part of the stress literature in fish more so than in mammals. This is mainly related to the fact that fish are exposed to aquatic pollutants via the extensive and delicate respiratory surface of the gills and, in seawater, also via drinking. The high bioavailability of many chemicals in water is an additional factor.

Together with the variety of highly sensitive perceptive mechanisms in the integument, this may explain why so many pollutants evoke an integrated stress response in fish in addition to their toxic effects at the cell and tissue levels. Exposure to chemicals may also directly compromise the stress response by interfering with specific neuroendocrine control mechanisms. Because hydromineral disturbance is inherent to stress in fish, external factors such as water pH, mineral composition, and ionic calcium levels have a significant impact on stressor intensity. Although the species studied comprise a small and nonrepresentative sample of the almost 20,000 known teleost species, there are many indications that the stress response is variable and flexible in fish, in line with the great diversity of adaptations that enable these animals to live in a large variety of aquatic habitats.

5

Hormonal Manipulation of Fishes

Aquaculture is the controlled reproduction and growing of aquatic organisms for food or ornament. Current trends in human population growth, habitat and natural stocks destruction versus advances in world agricultural and fisheries production clearly show the basic need for developing and implementing aquacultural technology.

Widespread use of hypophysation (use of the pituitary gland, a part of the brain and endocrine system of humans and fishes) has permitted breeding of a number of species otherwise not spawned in confinement (e.g. kuhli loaches, Acanthopthalmus kuhlii, dolphinfish, Coryphaena hippurus). Many of the important cultured pet and food fishes fall into this category. Previously fingerling-stock had to be collected from the wild. This supply was seasonal and unpredictable. By manipulating hormonal and environmental factors seed can be produced all year long; reducing cost, disease, and incidence of trash fish (undesireable contaminants) and predators.

Hypophysation involves the injection of gonadotropins (hormones=chemical messengers produced by the pituitary, that stimulate the reproductive organs) from one animal into another. Among vertebrates (fishes, amphibians, reptiles, birds and mammals) these are not very species-specific. Sometimes it has been found better to mix mammalian pituitary extracts and steroids in varying concentrations with fish gonadotropin. There is increased success in the use of prepared mammalian hormones in breeding procedures, preservation of fish sperm, maintenance of brood stock; yielding reliable sources of fish seed and acceleration of breeding research.

Problems relating to standardization of dosage, determination of maturity of recipient fish and nutritional requirements particularly need to be further explored. Current knowledge on the use of pituitary hormones in fish culture and technology regarding collecting, processing & storing of glands & methods of application have been covered comprehensively in the literature. The first experiments regarding reproductive behavior (in female mice) and replacement therapy using pituitary implants. The first use in fishes was in 1930 by B.A. Houssay of Argentina. He induced premature birth in a viviparous (livebearing, like guppies) fish by injecting pituitary glands recently removed from other fish. Other South Americans, like von Ihering, a Brazilian, used this technique for aquaculture extensively beginning in 1934.

Soviets attempted to induce captive sturgeons to spawn in 1932 by means of mammalian hormones. They were unsuccessful until 1937, when N.L. Gerbil'skii produced eggs and sperm in a number of the sturgeon, Acipenser stellatus that had been intracranially injected with 1-2 fresh pituitaries of conspecifics (the same species). As of 1957 the Soviets have obtained all their sturgeon eggs for culture from pituitary treated fishes, very important to their caviar industry. As of 1963 the Chinese have been able to supply all the seed used in carp culture (big-head, silver, common & other food species) by means of hypophysation.

Methods of Induced Breeding

Fish are induced to spawn by altering environmental parameters, particularly temperature and photoperiod, and hormonal constitution. These variables are generally constant for the species but vary widely among cultured food fishes.

Environmental Factors and Diet

Environment and diet are the principal sources of phenomena determining hormonal secretion. Several factors, photoperiod, temperature, metabolites, pheromones, light strength and temperature shock control release of gonadotrophins, which in turn induce gamete (eggs & sperm) production and concomitant reproductive behavior. Reproduction may occur out of season by manipulation of temperature and photoperiod (how long the light is on daily) alone.

An example of improved culture just through manipulation of environmental factors may illustrate the point. The Ayu, *Plecoglossus altivelus*, a smelt-like fish cultured in Japan during the summer months, normally spawns in October-November., after which most of the fish die. Due to lower water temperatures that time of year, it's difficult to maintain live food organisms in adequate densities in rearing ponds to feed fry. Adjustment of photoperiod resulted in successful acceleration of sexual maturation (August), enabling rearing of fry when natural food was abundant. Spawning was also delayed to prolong the life of adults. The light period was extended to 18 hours per day with artificial light (August through October). Reproduction was retarded & adults were marketed out of season in February.

Temperature

The relative importance of temperature varies per species, investigator and experiment, but all agree a certain range and lowering or elevating optimizes results. Smiglieski (1975) stated that temperature was the controlling factor in his work with flounders. Yamamoto et. al (1966) observed spermatogenesis is goldfish, Carassius auratus below 14 degrees C & accelerated spermiation above 20 degrees C. Often, if much temperature fluctuation occurs, reproductive behavior will cease. Haydock (1971) has observed temperature threshold below which gulf croaker will not hydrate or ovulate. For each species or sometimes race (a sub-specific classification) there are limits, both high and low, which eliminate reproduction physiologically and behaviorally.

Light

As with temperature, there are optimal amounts of strength, quality and duration of light in relation to reproductive behavior. On work with bluegill sunfishes it has been found that a longer photoperiod 16light/8dark versus 8L/16D induced a better gonosomatic index (G.S.I.= ovary weight divided by total body weight times 100). Males were more aggressive, dug more nests and fertilized more eggs.

Hoar (1965b) sums up the effect of longer light periods by stating that they stimulate secretory activity of anterior pituitary, inducing pre-sexual behavior. The pituitary LH, Lutenizing Hormone (one of the gonadotropic hormones produced in the brain) activates interstitial tissue of the gonads which produce gonadal steroids that in turn dominate sexual phases, taking complete control during prenatal phases. Several investigators have shown that elevated temperature and prolonged photoperiod increased gonad maturation in the green sunfish, Lepomis cyanellus.

Other

Photoperiod and temperature have received most attention & are generally considered to be of greatest importance in inducement of sexual maturity and spawning. Other factors such as the effects of other environmental stimuli; meteorological (rain, floods, etc.) & water conditions (pH, Ammonia, $C0_2$, turbidity) on controlled breeding of food and ornamental fishes has been investigated. In several species, the presence of con- (the same species) and/or hetero- (other) specifics is important. In bluegills and some cichlids it has been found that when males are present, the most aggressive female has greaater ovarian development. Another application of environmental control is employed in mullet culture in Israel. By keeping the adults in freshwater ponds and not allowing access to the sea where they migrate to spawn, fish-culturists are able to extend the spawning season.

Diet

Researchers have demonstrated gonadal regression induced by restricted diets accompanied by an inversion (opposite) of the lasophil/acidophil (cell types distinquished by their staining characteristics) ratio of the mesoadenohypophyis (an anterior part of the pituitary gland in ours and fishes brains) and a reduction in gonadotrophin content. These conditions were reversed with adequate food. Some authors have stressed the interaction between photoperiod and diet and pointed out that optimum benefit from photoperiod adjustment can be over-rided by poor diets. Hmmm.

C. Hormonal Manipulation

Criteria: Biological and Chemical Assay Several methods are applied towards determining sex and sexual maturity of spawners. Externally there may be sexual dimorphism (structural differences between the sexes) in the species and/or differences that occur during sexual ripeness. Such is the case with the channel catfish, Ictalurus punctatus. Females assume a dropsical appearance with raised and reddened genitalia. The males develop swollen head musculature and well defined genital papilla.

A practical and precise way to measure oocyte (egg cell) development is through inserting a sterile polyethylene or glass catheter in the oviduct and removing egg samples. The eggs are cleared in successive concentrations of ethyl (grain, drinking) alcohol (30, 50, 70, 90, 100%) then xylene (an organic solvent) for 5 minutes each. Eggs are removed by catheter before each hormone injection to determine gonadal development.

Physiological parameters are also valuable as sources for assays. Elevated plasma (the fluid portion of blood, the same as for humans) calcium levels have been reported during gonad development in many species. Peterson & Shehadeh (1971) reported a four time increase in calcium and two time plasma copper in striped mullet, Mugil cephalus, under methyl testosterone treatment.

For examination, gonads are removed from the body cavity, weighed to 0.01 gram, labelled, wrapped in cheese cloth, fixed in 70% ethanol (ethyl alcohol) for 30 minutes and 100% for one hour. Xylene is used for a clearing agent. Infiltration and final embedding is done in fresh paraffin (wax). Typically microscope slides are produced by slicing these specimens on a tool called a microtome at 8 micron thickness. These tissue slides are stained with a compound called Delafields iron hematoxylin and counter stained (for contrast) with eosin. The specimen is permanently covered with a thin glass coverslip and permount.

When you examine these prepared sections under a microscope you can judge the sexual readiness oth your stock. In sexually immature males the lumen (opening) of the seminiferous tubules is unformed. Later primary and secondary spermatocytes develop in the lumen. Still later spermatids and spermatozoa (these are all developmental phases of male reproductive cells) are seen free in the lumen. After spawning the tubules are completely occluded (blocked) and testes become highly vascularized (serviced with blood vessels). In immature females only the oogonia & primary (previtellogenic) oocytes are present. At maaturity secondary oocytes distributed by oolemma and centripedal yolk precursor form in the cytoplasm (the fluid portion of cells). The pituitary may be examined in sample individuals. They are dissected and promptly fixed in Bouin's or Zenker's for 24 hours. They are subsequently passed through alcohols and stored in 70% ethanol, further processed and sectioned. In the channel catfish the adult pituitary double from one to two millimetre maximum length at sexual ripeness. The mesoadonhypophysis contains gonadotropic and somatotropic cells; the former enlarge in ripe specimens. In this way the mesoadenohypophysis becomes large during sexual maturity.

Maturity of Recipient Fish

Variation in a species in terms of gonad development is intensified in confinement. Variability of results is due to individual variation in sex or season, especially considering crude bioassay techniques. Effectiveness of hypophysation (injection with pituitary hormones) is dependent on the stage of reproductive development of recipients. Aquaculturists recommend that when applying other workers' data; that matching the dosage to the physiological state of the recipient can best be done by administering doses in graded series from above or below those recommended by other investigators. Almost always slightly greater success is achieved with the most intense regime (injecting less hormone more frequently).

If the species has not been spawned before they suggest continuous injections, every 24 hours, with a wide range of doses until spawning does occur. As has been mentioned, care must be exercised in assaying sexual readiness in spawners. Parameters generally adopted have frequently proven unreliable. An example of this is females with enlarged abdomens, reddish colouration & protrusion of the cloacal region may be due to engorgement of the intestine, or disease, even during the spawning season.

Variability in Pituitary Gonadotropin Related to Phylogeny, Sex & Season

In order to determine dosage, effect and success of treatment it is vital that fluctuations in gonadotropic activity related to phylogenetic specificity (basically how closely species are related to each other evolutionarily), sex and season be known for each species. Much controversy exists as to specificity in gonadotropin as it relates to vertebrate phylogeny. Work with homoplastic (derived from the same species) pituitaries has produced best results. Little correlation exists between phylogenetic affinity and effect of pituitary gonadotropins. Where distantly related donor-recipients are used the general practice has been to increase dosage.

The early literature on sex related differences in the gonadotropic activity of fish pituitaries has been outlined by Pickford & Atz (1957). Much of it and later reports are contradictory. It is more or less agreed that sex-related differences in pituitary extracts are usually minimised by the usual preparation of mixed extracts of both sexes in excess of threshold requirements. Seasonal fluctuations in gonadotrophic activity are not uncommon in almost all fish species examined. Pituitaries collected for hypophysation are almost always taken from mature fishes during spawning season to be used immediately or dried and prepared for later use. (Kind of hard on your breeders, eh?)

Use of Mammalian Hormones

Advantages of using mammalian hormones include:

(1) Reliable source, available all the time.

(2) Easily stored in lab and retain biological activity for a number of years.

(3) More uniform than crude pituitary preparations, permitting controlled experiments and repeatable results.

(4) Cost is about the same.

(5) Saves valuable gravid fish, especially where homoplastic pituitaries are used.

(6) Does away with the tedious task of collecting, preserving, processing & distributing pituitary material.

There is not much in the literature on the use of such hormones for practical breeding of ornamental fishes; quite a bit on food species. Several Indian workers have worked extensively with Heteropneustes and mammalian hypophysial hormones, placental gonadotrophins, gonadal hormones and corticosteroids finding that human chorionic gonadotropin (HCG) and some of the corticoids, especially deoxycorticosteroids could be used profitably to spawn food fishes as a substitute for pituitary extracts. Chorionic gonadotrophin was also found to be an effective ovulating agent in Clarias batrachus. Snead & Clemens (1959) established methods for spawning channel catfish, goldfish, and white crappie, Pomoxis annularis with injections of chorionic gonadotrophins. At the price of $5.65 per 10,000 international units (IU) of APL (Anterior Pituitary Like) they established a cost of 45 cents per pound of fish spawned, comparable to established costs of fresh pituitary preparations. Other reviews of successful use of mammalian hormones exist detailing one's effectiveness over another, acting singly and together.

Purified fish gonadotropins are available to the fish breeder, especially fractionated salmon pituitary. Much work has been done with these and food fishes; not much reported for pet fish.

Pituitary Extracts, Use and Preservation

In a fish, the pituitary is concerned with regulation of growth, pigmentation, adrenal cortex, thyroid gland and others, as well as reproduction. Cyclical changes in the pituitary are striking, showing increase and decrease in cyanophil cells involved with production of gonad stimulating hormones. The synergic action of growth hormone, thyrotropic hormone (stimulating the thyroid gland) & sex steroids is indicated by the better results obtained with crude rather than purified gonadotropic hormones.

The pituitary regulates both gametogenesis and steroidogenesis in fishes. Two seperate gonadotropins Follicle-stimulating hormone (FSH) and Lutenizing hormone (LH) are physiologically distinct in the tetrapods (= "four legs", the rest of the vertebrates beyond the fishes, i.e. the amphibians, reptiles, birds & mammals). There is evidence of a single protein in fishes. When fish pituitary extracts are tested in tetrapods, both FSH and LH effects are elicited.

With rare exceptions induced spawning by hypophysation has been restricted to females. This is often due to the availability of male seed. Collection of pituitaries involves chopping the head off the fish above the eyes, lifting the brain and removing the gland from the base of the cranium. Preserving the gland is done by freezing immediately with dry ice or dehydrating in acetone. This is done by changing the acetone several times every 12 hours. This is then stored in a desiccator (dryer). It lasts up to two years but the effect is not as good as fresh.

When preparing, care must be taken to not introduce foreign proteins. Combined with pituitary this sometimes causes very strong rejection. Therefore purification is necessary. Whole fish pituitaries may be ground up or injected whole. If ground, they are suspended in physiological saline or distilled water.

Materials and Methods

Nomenclature of GnRH Ligands and Receptors: GnRH1 denotes the peptide that regulates reproduction in vertebrates, and we use mGnRH1 to distinguish the mammalian form. For other GnRH peptide sequences, we show amino acid changes relative to the mGnRH1 sequence (35). GnRH peptides used in this study were mGnRH1 (pGlu-His-Trp-Ser-Tyr-Gly-Leu-Arg-Pro-Gly-NH_2) and the three forms of GnRH found in *A. burtoni*, [Ser^8] GnRH1 (also called seabream GnRH), GnRH2 ([His^5,Trp^7,Tyr^8] GnRH) (36), and GnRH3 ([Trp^7,Leu^8] GnRH, also called salmon GnRH) (35, 37). GnRH-Rs are designated as in GenBank with the addition of a three-letter superscript, which is the conserved motif in EC3 (SHS, PEY, PPS, or SDP) to indicate their GnRH-R type classification. Hence, the receptor identified by our experiments becomes GnRH-$R1^{SHS}$, because it was designated GnRH-R1 in GenBank and contains the SHS sequence in EC3. Correspondingly, the other *A. burtoni* GnRH-R is named GnRH-$R2^{PEY}$. In receptors where the EC3 motif is not identical to the conserved motif, the motif in the receptor is used (*e.g.* GnRH-$R1^{PDY}$ for one of the catfish receptors, which has PDY instead of PEY in EC3).

Animals

A. (Haplochromis) burtoni derived from a wild-caught population were raised in aquaria under conditions similar to their native equatorial habitat in Lake Tanganyika in Africa (pH 8, 28 C) (28).

Both male and female fish (3–4 cm length) were used in these experiments. To collect brain tissue, fish were killed via rapid cervical transection. All work was performed in compliance with Stanford University (APLAC) guidelines.

Functional Assay of GnRH-R1^{SHS}

For functional tests, the *A. burtoni* GnRH-R1^{SHS} cDNA expression plasmid was transfected into COS-1 cells using diethylaminoethyl dextran, as previously described (40). The ligand structure-function relationships of the putative GnRH-R1^{SHS} were directly compared with *A. burtoni* GnRH-R2^{PEY} (32) by measuring ligand-stimulated inositol phosphate (IP) production, as described previously (40), except that IP was extracted with formic acid (10 mm). Ligand concentrations ranged from 10"^{10} to 10"^{5} m. Data points were determined in duplicate, and EC_{50} values are the means of three separate experiments.

For competition binding assays, [His^{5},d-Tyr^{6}] GnRH1 was radioiodinated as previously described (41). Transfected cells were washed twice with cold PBS and incubated (5 h at 4 C) with ^{125}I-labelled [His^{5},d-Tyr^{6}] GnRH1 (10^{5} cpm) in the absence (B0) or presence of various concentrations (10"^{9} to 10"^{5} m) of unlabelled peptides in 0.5 ml HEPES-buffered DMEM with BSA (0.1%). Unbound ^{125}I-labelled [His^{5},d-Tyr^{6}]-GnRH was removed by washing twice with PBS, and cell-associated radioactivity was solubilized in 1 ml 1 m NaOH and counted. Nonspecific binding was determined in the presence of excess GnRH2 (10"^{5} m). Data points were determined in duplicate, and the IC_{50} values are the means of three separate experiments.

Localisation of GnRH-R1^{SHS} mRNA Transcripts

GnRH-R1^{SHS} expression was localised using PCR on tissue samples from adult *A. burtoni* (brain, pituitary gland, retina, gill, heart, intestine, kidney, liver, muscle, ovary, retina, spleen, and testes) and *in situ* hybridization using antisense receptor-specific probes in the brain and pituitary.

In Situ Hybridization

In situ hybridization methods are described in detail elsewhere (33), so only essential details are given here. Reproductively active males (n = 3) were killed by rapid cervical transection, and brains were removed and placed in sterile PBS. Tissue was frozen in mounting medium (OCT; Tissue-Tek, Torrance, CA) on dry ice, sectioned coronally, and stored at “80 C until use. Probes for GnRH-R1^{SHS} and GnRH-R2^{PEY} were synthesized, based on GenBank sequences (GnRH-

R1^{SHS}, GenBank AY705931; GnRH-R2^{PEY}, AY028476), and plasmids were linearized to generate both sense and antisense templates for each receptor. GnRH-R1^{SHS} and GnRH-R2^{PEY} probes were from 151-1035 bp and from 55–1129 bp, respectively, and were generated by transcribing with SP6 or T7 polymerase in the presence of [^{35}S]UTP (Maxiscript kit; Ambion, Austin, TX). Slides with brain tissue were treated as described previously (33). Photomicrographs were captured digitally and optimized for clarity.

Regulation of GnRH-R mRNA Abundance

Tissue Preparation.

To discover whether expression of GnRH-R mRNA was regulated, either as a function of diurnal time or with respect to GnRH ligand expression patterns, animals (n = 24) were housed in separate aquaria with three animals of mixed sex including one T male in each of eight tanks to provide minimal disturbance of other animals in the room when the animals were killed. Tissue was collected from animals in a single tank at 3-h intervals over a 24-h period beginning at 1000 h (fish were maintained on a 12-h light, 12-h dark cycle with dark-to-light transitions at 0900 h). At night, the aquarium room was accessed through a light-tight door, and all tissue-processing steps were performed in complete darkness using minimal infrared illumination. Fish were killed by rapid cervical transection, and their brains and pituitaries were rapidly removed. Brain and pituitary samples were stored at "80 C until use.

Tissue taken at each time point was separately homogenized in chilled TRIzol (Invitrogen) with a Tissue Tearor and kept frozen in TRIzol reagent at "80 C until RNA extraction.

RNA Extraction and PCR Sample Preparation.

Total RNA was extracted from samples following a standard protocol (TRIzol; Invitrogen). RNA was DNase treated to remove genomic DNA contamination (TURBO DNA-free; Ambion), and 1.0 μg total RNA from each tissue sample was reverse transcribed to cDNA (SuperScript II RNase H reverse transcriptase; Invitrogen).

Quantitative Real-time PCR of A. Burtoni GnRH Ligands and Receptors

Primers for quantitative real-time PCR for [Ser^{8}] GnRH1 were 52 -cagacacactgggcaatatg-32 and 52 -ggccacactcgcaaga-32 , generating a 128-bp product; for GnRH2, primers were 52 -tggactcctttggcacatcagaga-32 and 52 -ctctggctaaggcatccagaagaa-32 , generating a 126-bp product; for GnRH3, 52 -atggatggctaccaggtggaaaga-32 and 52 -tggatttgggcatttgcctcatcg-32, generating a 11- bp product; for GnRH-R1^{SHS}, 52 -tcagtacagcggcgaaag-32 and 52 -gcatctacgggcatcacgat-32 , generating a 187-bp product; and for GnRH-R2^{PEY}, 52 -ggctgctcagttccgagtt-32 and 52 -cgcatcaccaccataccact-32 , generating a 220-bp product. Three housekeeping genes, actin, 18S rRNA, and tubulin, cloned in *A. burtoni* were used to control for sample differences in total cDNA content. PCR were performed (iCycler; Bio-Rad, Hercules, CA), and reaction progress in 30-µl reaction volumes was monitored by fluorescence detection at 490 nm during each annealing step. Reactions contained 2× IQ SYBR Green SuperMix (Bio-Rad), 10 µm of each primer, and 1 ng cDNA (RNA equivalent). Reaction conditions were 1 min at 95 C and then 35 cycles of 30 sec at 95 C, 30 sec at 60 C, and 30 sec at 72 C, followed by a melting curve analysis over the temperature range from 95 C to 4 C. All reactions were run in duplicate.

Quantitative Real-time PCR Data Analysis.

Fluorescence readings for each sample were baseline subtracted, and suitable fluorescence thresholds were automatically determined by the MyiQ software. To determine the number of cycles needed to reach threshold, the original fluorescence reading data were analyzed using a special PCR algorithm (43). To calculate the relative mRNA amount, target gene levels were normalised using the geometric mean of housekeeping gene mRNA values.

Statistical Analyses

To test the hypothesis that GnRH ligands and/or receptors were expressed with a diurnal rhythm, we analyzed the mRNA samples taken around the clock for rhythmic variation using generalised additive models (44, 45). This technique identifies nonlinear effects that might produce rhythmicity by analyzing the residuals of a scatterplot smoother fit to the data with a back-fitting algorithm. For the generalised additive models and the linear regressions, we took $P < 0.05$ as the initial significance level and applied a Bonferroni step-down correction to account for family-wise error due to multiple testing. These analyses were computed using R (R:Statistical computing environment; R Foundation, Vienna, Austria).

To identify potential linear relationships between ligands/receptors in a larger data set, we used robust regression (Siegel repeated means) (46) that allowed for the observed nonconstant variance (heteroskedasticity) in the data without requiring data transformation.

As a control, we similarly compared the relationships between the ligands/receptors and the Clock 1a gene (GenBank DQ923857). To look for independent sources underlying the data distribution, we used independent component analysis (ICA), which is a recently developed technique related to latent variable and factor analyses. Unlike more traditional approaches, ICA yields unique (except for the sign) solutions to the problem of finding underlying sources of variation by maximising the information contained in each independent source of interaction (44). The resulting sources may thus be interpreted as latent variables containing information about independent sources of variation in the data. We examined the relationships of each of the ligands/receptors with these latent variables by regressing each of the ligands/receptors on each of the independent sources, again using robust regression. We used a consensus source matrix from 100 solutions arrived at using the maximum-likelihood ICA package (computed using R). It is important to note that because the ICA signs are arbitrary, we can discover whether ligands/receptors relate to sources in the same or opposite directions or not at all (*i.e.* the signs can be reversed, but not the relative relationships between variables).

Results

Cloning of Full-Length GnRH-R1^{SHS} from A. Burtoni: Analysis of the full-length putative GnRH-R cDNA sequence from *A. burtoni* tissue shows it to be a second GnRH-R, with conserved Asp residues in transmembrane segments 2 and 7 and the carboxyl-terminal tail that are characteristic of GnRH-Rs that are not mammalian SDP-type GnRH-Rs (47, 48). There are conserved motifs SCAFVT in transmembrane segment 3 and VSHS in EC3 characteristic of teleost receptors as identified by Lethimonier *et al.* (12). This sequence is most similar to other SHS-type GnRH-Rs especially from teleost fish (*e.g.* GnRH-R1^{SHS} *O. niloticus*, GenBank AY705931). Consistent with our proposed nomenclature, we named this receptor GnRH-R1^{SHS} (12, 33, 34). The sequence also has residues conserved among all GnRH-Rs, Asp^{107}, Trp^{110}, Asn^{111}, Lys^{130}, Asn^{217}, and Tyr^{298}, which are equivalent to residues Asp^{98}, Trp^{101}, Asn^{102}, Lys^{121}, Asn^{212}, and Tyr^{290} of the human GnRH-R1^{SDP}, known to be important in ligand binding (8).

Function of the A. Burtoni GnRH-R1^{SHS}

To discover whether the *A. burtoni* GnRH-R1^{SHS} cDNA encoded a functional receptor, we analyzed its function using a recombinant system. Because fish cell lines were not available, we transfected COS-1 cells with GnRH-R1^{SHS} and showed specific binding of a radiolabelled GnRH analog, which was displaced by mGnRH1 and each of the three forms of GnRH expressed in *A. burtoni*. We found that GnRH-R1^{SHS} had relatively low affinity for the three forms of GnRH that are expressed in *A. burtoni*. IC_{50} values ranged from 53.5 nm for GnRH2, which had highest affinity for GnRH-R1^{SHS}, to 20,300 nm for [Ser^8] GnRH1. Consistent with our previous report, GnRH-R2^{PEY} exhibited similar low affinities for all peptides, and peptides had the same rank order of IC_{50} values in cells expressing either GnRH-R1^{SHS} or GnRH-R2^{PEY} (GnRH2 < GnRH3 < mGnRH1 < [Ser^8] GnRH1).

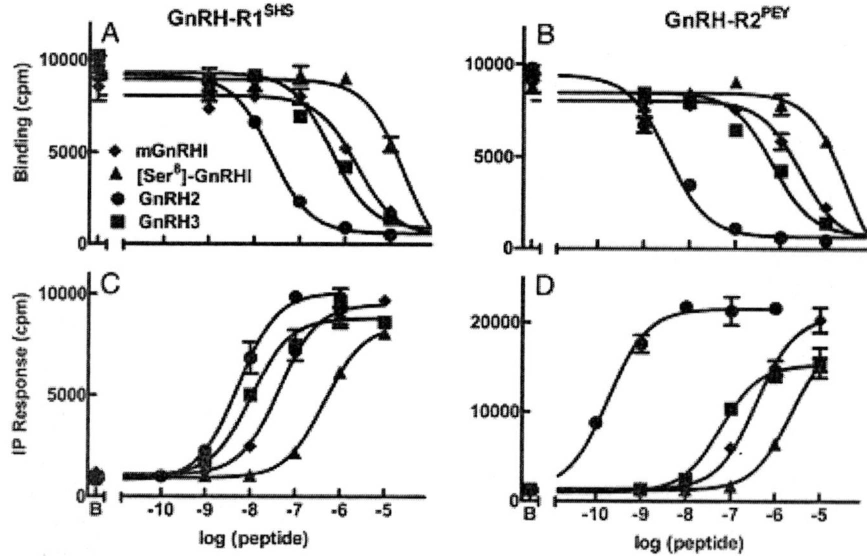

Figure: Competition binding and IP production of GnRH-R1^{SHS} and GnRH-R2^{PEY} receptors. COS-1 cells transiently transfected with cichlid GnRH-R1^{SHS} (A and C) or cichlid GnRH-R2^{PEY} (B and D) were incubated with ^{125}I-labelled [His^5,d-Tyr^6] GnRH (A and B) in the presence of increasing concentrations of mGnRH1 (f&), [Ser^8] GnRH1 (´%), GnRH2 (•), or GnRH3 (^{a}%) for competition binding assays or (C and D) labelled with [^{3}H]myoinositol and incubated with increasing concentrations of peptides in the presence of LiCl for peptide-stimulated IP production.

The four GnRH peptides stimulated IP production in cells transfected with *A. burtoni* GnRH-R1^{SHS} cDNA, confirming that the cDNA encodes a functional receptor that activates the Gq/11 family of G proteins in COS-1 cells and couples to the IP second-messenger cascade. GnRH2 and GnRH3 exhibited the highest potency (EC_{50} values 8.11 ± 2.63 and 11.0 ± 2.5 nm, respectively), whereas [Ser^8]

GnRH1 and mGnRH1 had low potency. In contrast, only GnRH2 exhibited high potency at the GnRH-R2^{PEY} (EC_{50} 0.208 ± 0.034 nm).

In vitro analyses show that the *A. burtoni* GnRH-Rs have similar affinities for each of the GnRH forms in *A. burtoni.* However, comparison of IC_{50} and EC_{50} values indicates that the two receptors differ in their efficiency in coupling binding of different GnRH peptides to intracellular signaling. This is evident from the similar potencies of GnRH2 and GnRH3 (EC_{50} values 8.11 ± 2.63 and 11.0 ± 2.5 nm) at the GnRH-R1^{SHS}, even though GnRH3 has considerably lower affinity (IC_{50}, 1143 ± 578 nm) at this receptor than GnRH2 (IC_{50}, 53.5 ± 24.6 nm). To quantify these observations, we have calculated an efficiency quotient (IC_{50}/EC_{50}, Table 1Ñ!). Clearly, the GnRH-R1^{SHS} signaling response is particularly sensitive to GnRH3 and poorly responsive to GnRH2, whereas the GnRH-R2^{PEY} is highly responsive to GnRH2. It is notable that neither receptor is highly responsive to [Ser^{8}] GnRH1.

Tissue Distribution and Localisation of A. Burtoni GnRH-R1^{SHS} and GnRH-R2^{PEY}

We mapped the distribution of both GnRH-Rs in *A. burtoni* using RT-PCR for peripheral tissue and *in situ* hybridization for expression in the brain. GnRH-R1^{SHS} is significantly more widespread in the body than is GnRH-R2^{PEY}, being expressed in muscle, intestine, liver, and heart in addition to brain, retina, pituitary, gill, kidney, testes, and ovary where both forms are expressed.

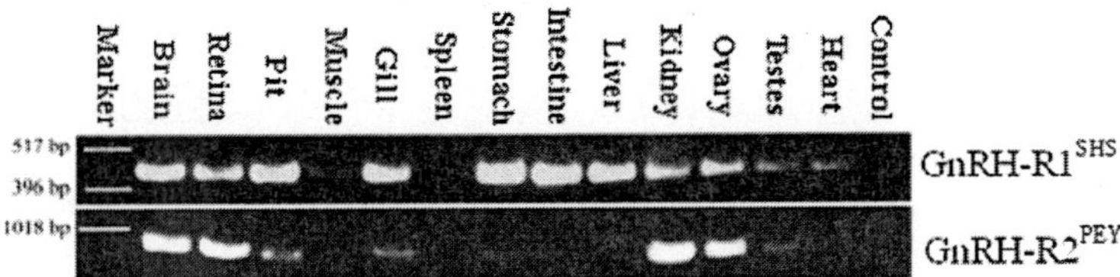

Figure: *Tissue distribution of GnRH-Rs. Distribution of mRNA from both GnRH-R types in* A. burtoni *tissue was assessed using PCR. Note that both receptor types are distributed throughout the body, but GnRH-R1^{SHS} is found more widely than GnRH-R2^{PEY}, particularly in the muscle, intestine, liver, and heart. Neither form is found in spleen. The first lane contains markers for 517 and 396 bp for GnRH-R1^{SHS} and 1018 bp for GnRH-R2^{PEY}. Material for the control lane was reverse transcribed without RNA and then used as the PCR template.*

To localise GnRH-Rs relative to the gonadotropes and somatotropes, we compared GnRH cell type locations visualized using immunocytochemistry with each receptor type mapped using *in situ* hybridization. The organisation of the teleost pituitary requires that it be sectioned parasagittally to see these relationships clearly. In sagittal sections of the pituitary, the distributions of cells containing GH or LH and a schematic representation are shown. GH- and LH-containing cells are in nonoverlapping regions of the pituitary. Distribution of FSH-containing cells was similar to that of LH-containing cells (not shown).

Based on *in situ* hybridization, GnRH-R1^{SHS} is localised in a crescent around the ventral border of the pituitary, and GnRH-R2^{PEY} is located dorsally, directly adjacent to the attachment of the pituitary to the brain, confirming previous results for *A. burtoni* (33). Comparing the *in situ* hybridization results with those of GH and LH distribution in the pituitary of *A. burtoni*, it is clear that the distribution of the GnRH-R1^{SHS} correlates with gonadotropes and GnRH-R2^{PEY} with somatotropes.

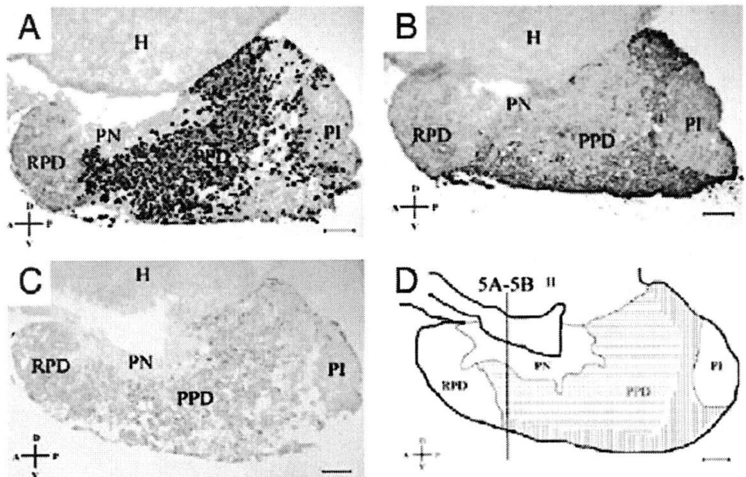

***Figure:** Localisation of gonadotropes and somatotropes.*

Fixed sagittal sections of *A. burtoni* pituitary from T males were incubated with anti-GH antibody (A) anti-LH (B), or no antibody (C). Bound antibodies were visualized with 32 ,32 -diaminobenzidine, and nuclei were counterstained with cresyl violet. A schematic illustration (D) shows the distribution of GH cells (*horizontal shading*) and LH cells (*vertical shading*). The *line* labelled 5A-5B shows the position of the coronal sections of *in situ* staining. *Scale bar*, 100 mm. A, Anterior; D, dorsal; P, posterior; V, ventral; H, hypothalamus; RPD, rostral pars distalis; PN, pars nervosa; PI, pars intermedia.

Photomicrographs show coronal sections through an *A. burtoni* pituitary showing localisation of GnRH-R2^{PEY} mRNA (A) and GnRH-

R1^{SHS} mRNA (B). A1, GnRH-R2^{PEY} mRNA as radioactively labelled silver grains (*black spots*) in bright field with cresyl violet counterstain (*purple*), showing that it is concentrated in the dorsal part of the pituitary in the proximal pars distalis (PPD) adjacent to the pituitary stalk; A2, dark-field photomicrograph of the same field of view where silver grains appear *silver*.

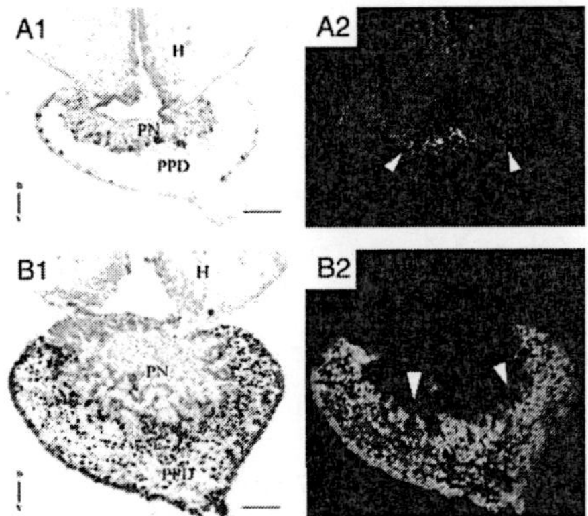

Figure:*Localisation of GnRH-R1^{SHS} and GnRH-R2^{PEY} in the pituitary of* A. burtoni *using* in situ *hybridization with ^{35}S-labelled antisense mRNA probes.*

Arrowheads mark the boundary of the staining that lies between the proximal pars distalis (PPD) and the pars nervosa (PN). B1, GnRH-R1^{SHS} mRNA as silver grains (*black spots*) in bright field with cresyl violet counterstain (*purple*), showing that it is concentrated in ventral crescent of the PPD; B2, dark-field photomicrograph of the same field of view where silver grains appear *silver*. *Arrowheads* mark the boundary of the staining that lies along the ventral border of the pituitary. *Scale bar*, 200 mm.

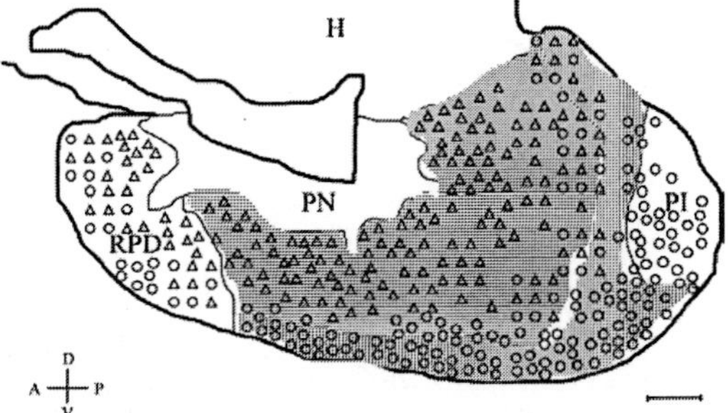

Figure: *Schematic illustration of a sagittal section through the* A. burtoni *pituitary showing the distribution of GnRH-R1^{SHS} (Ë%) and GnRH-R2^{PEY} (μ%) receptor types relative to the GH cells (*horizontal lines*) and LH cells (*vertical lines*).* Scale bar, *100 μm.*

Diurnal Rhythm and Coregulation of GnRH-R mRNA Levels

In *A. burtoni*, reproductive behavior in T males and activity in non-T males show distinct daily rhythms (28).

To determine whether there might be a corresponding rhythm at the GnRH-R mRNA level, we collected tissue from three animals at 3-h intervals over a 24-h period and performed quantitative RT-PCR to determine the levels of receptor expression. Although previous reports suggested GnRH-R mRNA levels change with a diurnal rhythm, we found no statistically significant nonlinearity as would be consistent with a diurnal rhythm in either GnRH-R1^{SHS} or GnRH-R2^{PEY} expression.

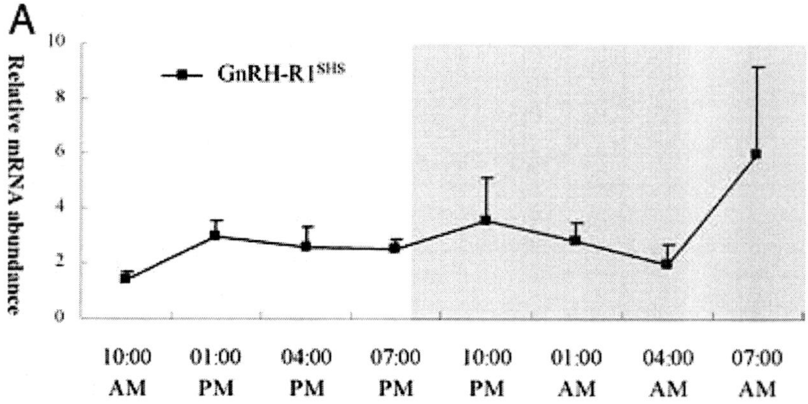

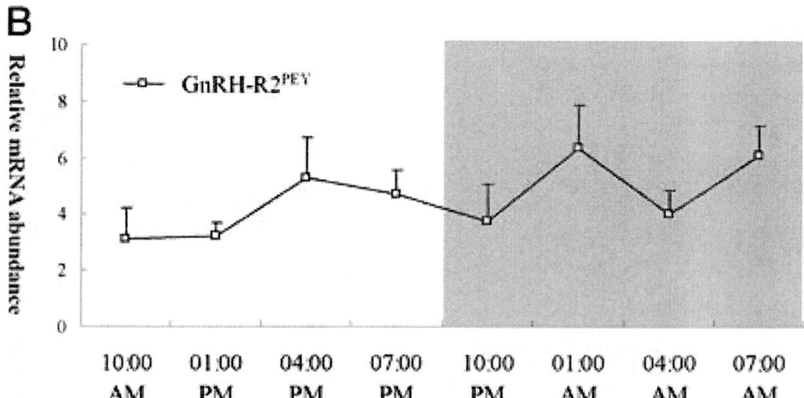

Figure: *GnRH-R mRNA abundance as a function of time of day over a 24-h period. Panels show the fluctuation of relative GnRH-R1*^{SHS} *(A) and GnRH-R2*^{PEY} *(B) mRNA abundance over 24 h.* A. burtoni *tissue samples were taken at 3-h intervals. mRNA abundance was normalised by mRNA levels of housekeeping genes.* Gray shaded area *represents darkness.* Bars *indicate sem. Lights were on at 0900 h and off at 2100 h (n = 24).*

However, we did find significant coregulation of particular forms of GnRH and GnRH-R mRNA in data collected around the clock from the brain and pituitary. There is a highly significant positive linear relationship between each receptor and all three ligands. ICA showed that two independent sources contribute to changes in all the variables in the same direction. This means that when values in the first underlying source's distribution increase, expression of all the mRNAs increases. A third underlying source, however, shows GnRH-R2^{PEY} changing opposite of all other variables. Interestingly, the third underlying source shows GnRH3 is expressed in the same direction as receptor 2 but in the opposite direction of GnRH-R1^{SHS}.

Phylogenetic Relationships of Jawed Vertebrate GnRH-Rs

Phylogenetic analysis identified four GnRH-R subfamilies with characteristic motifs in EC3. We have defined GnRH-R types according to their EC3 motif, indicated by a three-letter superscript, to reduce confusion of multiple contradictory nomenclature systems for GnRH-Rs and poor correlation of assigned receptor names with receptor type. This unambiguous classification scheme placing GnRH-Rs into four types, SDP, PEY, PPS, and SHS using the EC3 motif, is based solely on receptor amino acid sequence. It has major advantages in that it includes all but a few closely related SHS- and PPS-type receptors and cannot be confused with GenBank assigned receptor names.

It is likely that the four GnRH-R subfamilies identified (a1, a2, b1, and b2, corresponding to types PEY, SDP, PPS, and SHS) arose through gene duplication, an important process in evolution. Duplications allow new gene functions to evolve, and there is good evidence in yeast that duplicated genes can provide functional compensation against null mutations. In this case, GnRH-R^{SHS} and GnRH-R^{PEY} show similar affinity for [Ser^{8}] GnRH1, but they are clearly not functionally redundant, because their expression patterns are divergent.

GnRH-R subfamilies can be understood in terms of the two rounds of genome duplication that occurred early in the vertebrate lineage (known as the 2R hypothesis) (53). Because this hypothesis would predict the presence of four GnRH-Rs in all vertebrates, the phylogenetic tree predicts that GnRH-R genes have been lost independently, on multiple occasions from different branches of the vertebrate lineages. For example, SDP-type (family a2) GnRH-Rs have been identified only in mammals and thus must have been lost independently in fish, amphibians, and reptiles. Similarly, PEY-type (family a1) GnRH-Rs are not found in mammals. Amphibians have GnRH-Rs in both subfamilies of family b (PPS-type and P/SQS variations of the SHS type), whereas all fish GnRH-Rs in family b are SHS type and all bird and mammalian receptors are PPS type. Although GnRH-R^{PPS} receptors have been identified in amphibians, chicken, and nonhuman primates, this subtype is absent from the mouse and rat genome sequence data bases. The human GnRH-RII^{PPS} lacks a start codon, is truncated at the second transmembrane segment, and is nonfunctional (54, 55, 56). Similarly, a sheep GnRH-RII^{PPS} ortholog is also nonfunctional (57). Thus, this receptor has been independently lost in rodents and has become a pseudogene in humans and sheep.

Teleost fish have undergone an additional genome duplication since their divergence from tetrapods (58, 59), so it is not unexpected that there are multiple GnRH-Rs of a single receptor type in the *Ostariophysi* lineage including goldfish (GnRH-RA^{PEY} and GnRH-RB^{PEY}) (38), catfish (GnRH-$R1^{PDY}$ and GnRH-$R2^{PEY}$) (25), and zebrafish (GnRH-$R1^{PEY}$, GnRH-$R3^{PEY}$, GnRH-$R2^{SHS}$, and GnRH-$R4^{SHS}$) (12). Similarly, in the *Neoteleostei* lineage multiple GnRH-Rs are present in pufferfish (GnRH-R1/III-1^{SHS}, GnRH-R1/III-2^{SHA}, and GnRH-R1/III-3^{SHS}) and medaka (GnRH-$R1^{SHS}$ and GnRH-$R3^{SHS}$).

In the case of *A. burtoni*, the different expression patterns of the GnRH-R genes suggest that these receptors are in the process of diverging in function. Only GnRH-$R2^{PEY}$ is colocalised on all three groups of neurons that produce GnRH, apparently specialised for feedback control of GnRH production (33). Moreover, GnRH-$R1^{SHS}$ colocalises with gonadotropes, whereas GnRH-$R2^{PEY}$ colocalises with somatotropes, revealing different roles in the pituitary. When detailed data about receptor localisation and function are available from other species, it will become clear whether these specialisations are unique to teleosts and whether evolution has taken different directions in different species.

Role of GnRH-$R1^{SHS}$ as a Functional Receptor for [Ser^8] GnRH1

We have previously shown, in *A. burtoni*, that [Ser^8] GnRH1 is the only GnRH peptide found in the pituitary and that it is up-regulated in reproductively active T males. The low affinity of both *A. burtoni* receptors for [Ser^8] GnRH1 indicates that high concentrations of [Ser^8] GnRH1 (in the micromolar range) would be required in the pituitary to activate either receptor. Superficially, this might suggest that neither receptor is the target of [Ser^8] GnRH1 but that some other receptor with a higher affinity for [Ser^8]-GnRH1 may exist.

However, all of the nonmammalian GnRH-Rs that have been functionally analyzed have the highest affinity for GnRH2, and all fish receptors are poorly responsive to GnRH1 forms (11, 12, 25, 27, 38). Available information from zebrafish and pufferfish genomes does not reveal additional GnRH-Rs, making it unlikely that receptors with high similarity to currently known GnRH receptors and high affinity for GnRH1 exist.

Although [Ser^8] GnRH1 had low potency in stimulating IP production in our recombinant system, the efficiency of [Ser^8] GnRH1-stimulated activation of fish G proteins is not known. High-affinity GnRH-R binding has been reported in fish pituitary tissue (60, 61, 62, 63) using synthetic, high-affinity ligands, before the discovery of the teleost forms of GnRH1.

However, consistent with results in recombinant systems, catfish GnRH1 was found to have low potency in stimulating gonadotropin release and had very low affinity in GnRH-R binding assays performed with catfish pituitary membranes (60). This suggests the known receptors may regulate reproduction despite their low affinity for the biologically relevant ligands, raising the possibility that synchronous release of GnRH in the pituitary may lead to transiently elevated levels triggering the receptors.

Consistent with their known functions in mammals, all SDP-type GnRH-Rs are highly selective for mGnRH1, and all PPS-type receptors are highly selective for GnRH2. This suggests that PPS-type GnRH-Rs may have a conserved function in amphibians and in nonhuman primates. As in *A. burtoni*, all SHS-type GnRH-Rs exhibit relatively low ligand selectivity, with similar responses to different GnRH peptides, suggesting that SHS-type receptors may have similar functions in teleost and amphibian species.

In contrast, all PEY-type GnRH-Rs from fish species are highly selective for GnRH2, suggesting that GnRH2 may be their major physiological ligand. However, PEY-type receptors from chicken and bullfrog are much less selective for GnRH2 and are quite responsive to GnRH1 ([Gln^8] GnRH1 or mGnRH1, respectively) suggesting that the functions of PEY-type GnRH-Rs may differ between teleosts and other vertebrates. Indeed, there is physiological evidence that the GnRH-R1^{PEY} regulates reproduction in chickens, because its expression is up-regulated in the pituitaries of castrated cockerels. However, these data are not consistent with a recent study in which expression of GnRH-R2^{SHS} was shown to increase with reproductive status in male and female chickens (65).

Coregulation of GnRH Ligand and Receptor Expression

The widespread distribution of GnRH-Rs (66, 67) has led to speculation that different forms of GnRH may play roles beyond reproduction. Chen and Fernald recently showed that the receptor types differ in their brain distribution in *A. burtoni* and only GnRH-R2^{PEY} colocalised with the three GnRH-producing cell types in the brain, suggesting direct feedback control of the GnRH production. The coregulation of receptors with GnRH suggests that the GnRH system has some common regulatory underpinnings. The presence of multiple GnRH forms across all vertebrates suggests they might share some functional role in reproduction with GnRH1.

GnRH-R Evolution and the Usefulness of the EC3 Motif Classification System

The poor correlation of GnRH-R sequence with a role in reproduction across species makes it impossible to predict which GnRH-R will regulate reproductive function in any particular species. The absence of SHS-type GnRH-Rs in higher vertebrates precludes their role as the receptor responsible for reproduction in these species. Different families of duplicated receptor genes regulate reproductive function in mammals (family a, SDP-type) compared with fish (family b, SHS-type).

There is insufficient evidence in other taxa (birds, reptiles, and amphibians) to conclude which receptor is used or when selection for one receptor rather than another might have occurred. Thus, it makes sense to define GnRH-R types by genotype, which, perforce, correlates with EC3 amino acid motif and with ligand selectivity, rather than using a classification system based on physiological function in the face of insufficient and conflicting information.

In summary, we cloned the GnRH-R1^{SHS} receptor, which is up-regulated in reproductively active *A. burtoni*, and showed that it is coexpressed with LH in the pituitary. Like all teleost GnRH-Rs, it binds and responds poorly to GnRH1.

Although this remains puzzling, it suggests that synchronous release of GnRH1 into the pituitary allows delivery of sufficient ligand to activate the receptor. The absence of the SHS-type GnRH-R in many taxa and the poor correlation of GnRH-R sequence with a role in reproduction across species, suggest that different families of duplicated GnRH-R genes have been selected to regulate reproduction during vertebrate evolution.

Reproductive Physiology of Fish: a 30 Year Overview

Research into the physiology of fish has had many beneficial implications for the aquaculture industry, explains Bernard Jalabert, Fish Reproduction Group. This feature was published by Cybium, Société Française d'Ichtyologie.

Concluding Remarks Concerning the Application of Fish Reproductive Physiology Research to Aquaculture

Research in RPF represents a large scientific community all over the world, sustained by public, economic and environmental concerns in many countries. Scientific approaches concern various complementary levels, from basic knowledge to directly applied research.

In many cases, practical demand, originating mainly from the fish farming sector, has been pushing research to find solutions for the control of reproductive traits of economic interest, almost independently of research into underlying physiological regulations. This has been the case concerning manipulation of external factors, such as photoperiod or temperature. Thus, photoperiod manipulations have been applied in many farmed species to control out-of-season spawning (Bromage *et al.*, 2001) and puberty (Okuzawa, 2002). In both cases, however, precise protocols depend not only on the species but also on the genetic strain: in rainbow trout for example, the same photoperiod regime cannot be applied to an autumn-spawning and a spring-spawning strain. Moreover, interactions with other factors are either unknown or cannot be easily taken into account for practical reasons. This is the case for temperature, for which specific requirements are not limited to the spawning period, thus impairing egg quality, as shown in Arctic char (Gillet, 1991), rainbow trout (Davies and Bromage, 2002), Atlantic salmon (King *et al.*, 2007), or halibut (Brown *et al.*, 2006). Besides, inadequate photoperiod manipulation by itself can also impair egg quality, as in trout (Bonnet *et al.*, 2007) or in Eurasian perch (Migaud *et al.*, 006). In fact, the manipulation of external factors to control any reproductive step is generally limited by the ignorance of the precise mechanisms by which these factors act, and interact, on underlying physiological processes. As already underlined in the case of puberty, much can be expected from largescale and high-throughput methods applied to functional genomics of whole tissue transcriptome and proteome, to understand multiple interactions between external factors and physiological regulations.

Applications of RPF knowledge generally bring progressive improvements to aquaculture. This is the case, for example, of hormonal spawning induction. Initially a solution was provided in the form of injections of crude pituitary extracts, which were successively replaced by partially purified gonadotropin, then by various GnRH analogs, sometimes in combination with anti-dopaminergic compounds (reviewed by Zohar and Mylonas, 2001). However, rapid industrial development of aquaculture may follow on from the development of techniques that unlock a reproductive block, as in *Pangasius* fish farming in Vietnam. The annual farmed production of this species, based on fry capture in the wild, was stagnating at around 12500 tons until 1995. Then, the demonstration that successful maturation and ovulation could be artificially induced by serial hCG injections (Cacot *et al.*, 2002; Cacot and Lazard, 2004), boosted annual production up to 440000 tons in 2005 and probably near 1 million tons in 2007. Most of the increase concerns *Pangasius hypophthalmus* species, initially considered as less valuable with regards to flesh quality. This apparent paradox results from higher fertility and easier induction treatment of *Pangasius hypophthalmus*, but it also points to the essential limiting factors of species domestication, such as fertility (number of eggs per fish) and practicability of possible treatments in field conditions.

Finfish aquaculture based on seed capture in the wild, or even on the capture of adults, still concerns species such as eels, tunas, groupers and yellowtails, some of which are now considered as endangered species. Therefore, research on RPF has still much scope for progress, not only to remove the pressure on the species mentioned above, but also to find new candidate species for aquaculture by increasing efforts in the field of comparative reproductive biology of new species.

Finally, it may happen that advances in RPF knowledge open the door to innovative techniques that do not result of any demand from farmed fish producers and that can even be rejected by consumers in some countries, like transgenic fish for specific characteristics such as improved growth or disease resistance. In this field particularly, but also in the case of any innovation, public concern about biotechnological applications and sustainable aquaculture development must be taken into account.

Studies on the Reproductive and Developmental Biology of Cichlasoma Dimerus *(Percifomes, Cichlidae)*

ABSTRACT: Many characteristics of the South American teleost fish *Cichlasoma dimerus* (body size, easy breeding, undemanding maintenance) make it amenable to laboratory studies. In the last years, many of the fundamental aspects of its reproductive and developmental biology have been addressed in our laboratory. Rather recently, the immunohistochemical localisation of pituitary hormones involved in reproduction and in background colour adaptation has been described in both adult and developing individuals, and the role of FSH in ovarian differentiation has been established.

These findings have been correlated with mapping of some of their brain-derived controlling hormones. The latter include brain-derived gonadotropins which were shown to be active *in vitro* in the control of pituitary hormone secretions. The emerging picture shows *C. dimerus* as an interesting species in which many of their basic features have already been investigated and which conform a solid platform for comparative studies correlating neurohormones, pituitary hormones and behavior, from the molecular to the organismic level.

Teleost Fishes in Physiological and Developmental Studies.

Teleost fishes represent the most numerous and diversified vertebrate group. Though fish physiology has been studied for many decades (Ball and Baker, 1969; Hoar, 1969; Conlon, 2000; Price *et al.*, 2008), the use of very different species, together with the idea that differences were more important than similarities, gathered a lot of data, and sometimes non coherent ones. During the last years, however, several research groups have put the emphasis in the importance of using teleost fishes not for comparative studies within the class, but as general vertebrate physiological models, by taking advantage of some of their unique properties. One of these properties is an unusual characteristic of the fish nervous system: fish brains can remain alive and responsive for many hours after removal from the animal.

This has allowed researchers to use intactbrain preparations that have the advantage, compared to mammalian brain-slice preparations, of allowing afferent and efferent connections to remain functional (Wayne *et al.*, 2005). Another interesting feature is the morphological arrangement of fish neuroendocrine systems: in teleost fishes there is no median eminence or portal system; therefore nerve fibres originating from the hypothalamus should end in close contact with pituitary cells. Immunocytochemical studies have shown that neurohormonal fibres and nerve endings are in close association with pituitary endocrine cells. Because of this special anatomical arrangement, teleosts are a unique experimental model for determining putative brain peptides or monoamines involved in the regulation of pituitary endocrine cells (Peter *et al.*, 1990).

Also, over the past 15 years, the cyprinid zebrafish *Danio rerio* (Hamilton, 1822) and the adrianichthyid medaka *Oryzias latipes* (Temminck and Schlegel, 1846) have become very popular fish models, due to their transparency during embryonic and larval stages, for investigating the molecular genetics that control organ development. The by-product of this huge amount of work is that their genes have been characterised providing fundamental tools for physiological and developmental studies (Furutami-Seiki and Wittbrodt, 2004; Crollius and Weissenbach, 2009).

Cichlid fishes (Perciformes) are also becoming intensively studied in the field of social control of reproduction and the most studied species have been the African tilapia *Astatotilapia burtoni* (Günther, 1894) and *Oreochromis mossambicus* (Peters, 1858) (Fox *et al.*, 1997; Ogawa *et al.*, 2006). The South American cichlid fish *Cichlasoma dimerus* (Heckel, 1840) has been studied for several years in our laboratory, at the morphological, developmental, physiological, and molecular levels. The aim of the current review has been to present a comprehensive picture of all the data obtained in this species, and to put them in the context of contemporary, comparative fish physiology.

Cichlasoma dimerus in the laboratory and in nature.

It is a medium sized fish (12 cm standard length) which is easily maintained and bred in the laboratory, where it tolerates a wide range of water compositions and temperatures (10 to 30°C, the optimal breeding temperature is 26°C) (Meijide and Guerrero, 2000).

The natural range of this species encompasses the entire system of the Paraguay river, the lower Alto Paraná, and the rest of the Paraná river basin up to the vicinity of Buenos Aires city (Kullander, 1983). Locality records are known from four countries (Bolivia, Brazil, Paraguay and Argentina) where it inhabits a wide variety of both lentic and lotic environments. Its common names are "chanchita" (Spanish) and "acará" (Portuguese) (Staeck and Linke, 1995; Casciotta *et al.*, 2005).

Its ground colour is very variable, depending both on the specimen's "mood" and social status. Its body may vary from greenish to light or dark grey and it may also show golden-yellow and light blue reflections. It shows several dark-brown vertical bands and two blotches, one in the middle region of the trunk and the other in the caudal peduncle. The eyes may show a bright red border, particularly in reproductively active individuals. This species has a moderately developed sexual dimorphism, with the males growing larger than the females. In males, soft rays of the distal edge of the dorsal and anal fins may be extended as filaments (Alonso *et al.*, 2007).

Figure: *Some behavioural aspects of Cichlasoma dimerus under laboratory conditions. A: Fin display and body colours of an adult male of high social rank. B: Prespawning behavior: the female nibbles at the spawning substrate (a flat stone), while the male is on guard. C: Spawning behavior: the female rubs the genital papilla while depositing the last egg rows. D: Parental care of the clutch: the female is removing unfertilized or infected eggs. Scale bars = 20 mm.*

As many cichlids do, *C. dimerus* has highly organised breeding activities, as they may be observed in the laboratory. A few days after fish have been transferred to a new aquarium, territories are progressively established and defended by both males and females. The dominant pair will aggressively defend the prospective spawning site (usually a flat stone) and will start to display stereotyped prespawning activities, such as jerking and quivering, digging in the gravel, and nipping off and nibbling at the spawning substrate.

Spawning will eventually initiate, with the female rubbing her genital papilla over the substrate and the sticky eggs being deposited in rows. After the female lays its eggs, the male slides its erected genital papilla over them, thus achieving fertilization. At the end of the spawning process, which may last up to 90 min, some 1500 eggs have been deposited as a uniform rounded layer over the substrate.

Spawning is followed by a period of parental care during which the cooperative pair guards the eggs, both by fanning them (alternately beating the pectoral fins over the clutch) and by removing dead eggs, which are usually attacked by fungi. At a water temperature of 26°C, the larvae hatch at the beginning of the third day and are transferred by both parents to a previously dug pit. Another five days pass until the larvae swim freely, and during this time they are recurrently transferred to several other pits. Mortality during the first month of development is typically high (only about 20% of hatchlings reach the juvenile stage). If provided with enough food and space, the young will grow rapidly, and may reach about 50 mm four months after hatching. A pair may spawn every 20 days during the extended breeding season (September to April). The foregoing behavioral description was based on the works of Greenberg *et al.* (1965) and Polder (1971) in *Aequidens portalegrensis*, which is a synonym of *C. dimerus*. A comprehensive monograph on cichlid behavior by Baerends and Baerends-Van Roon (1950) is also available.

Gonadal Development and Sex Differentiation

C. dimerus is a gonochoristic fish, in which ovaries and testes develop directly from undifferentiated gonads. A detailed description of early gonadogenesis in *C. dimerus*, including ultrastructural features, is found in Meijide *et al.* (2005). As it is known for other teleosts (Auslé and Brusié, 1983), the extent of gonadal development in *C. dimerus* depends both on age and size, and gonadal sex differentiation will occur earlier in specimens that grow faster. Therefore, the ages corresponding to the developmental events described in the next paragraphs should be considered only as estimations.

On day 12 posthatching, paired gonadal primordia are present in the posterior region of the abdominal cavity, immediately below the kidney, and are suspended from the dorsal peritoneal wall by short mesenteries. The gonadal primordium consists of large, round to oval germ cells surrounded by somatic cells. Germ cells contain large and central euchromatic nuclei, with prominent nucleoli. Occasional mitotic activity is observed in these cells. Somatic cells are of varying shapes. They have heterochromatic nuclei and a small cytoplasm. Gonads remain undifferentiated by day 38. However, the number of germ cells has augmented, and as a result, gonads increase in size and several germ cells (not just one or two as in earlier stages) appear in cross sections.

Ovarian differentiation occurs at an earlier age than that of testicular differentiation, as it is usually observed in teleosts (Meijide *et al.*, 2005). A prominent cytological feature marking the onset of gonadal differentiation is the appearance of meiotic figures in germ cells of the prospective ovary. By day 42, many oogonia have entered meiotic prophase and become growing primary oocytes, which in addition to a moderate oogonial and somatic cell proliferation result in an enlargement of the prospective ovary.

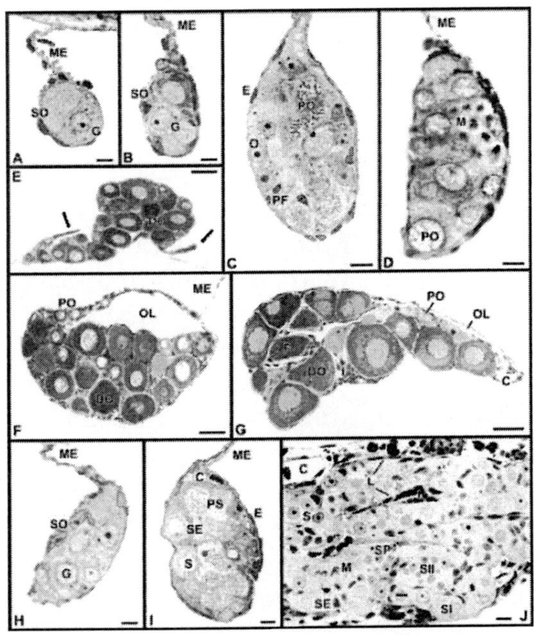

Figure: *Gonadal development and sex differentiation in Cichlasoma dimerus. Toluidine blue. A: Primordial gonads on day 14 post-fertilization are characterised by the occurrence of a few germ cells with round nuclei, which are surrounded by some somatic cells. B: Sexually undifferentiated gonad on day 38. C: The appearance of meiotic figures is the first evidence of sexual differentiation in prospective females on day 42. D: Increased gonial mitotic activity accompanies the first cells undergoing meiosis towards day 50. E: Arrows indicate somatic cell outgrowths in a female on day 50, which will lead to the formation of the ovarian cavity. F: The ovarian cavity is already formed on day 65. Numerous basophilic diplotene oocytes are also observed. G: Ovarian follicles containing perinucleolar oocytes are observed by day 80. H: An undifferentiated gonad on a day 65 juvenile (a prospective male). I: Meiotic figures indicative of testicular differentiation are first seen on day 80 in a prospective male. J: Lobules are becoming distinct in a testis on day 100. Secondary spermatocytes (but no spermatids) are observed. Scale bars = 5 μm (A, B, H, I); 10 μm (C, D, J); 50 μm (E, F, G).; C, blood capillary; DO, diplotene oocyte; E, epithelial cell; F, follicle cell; G, germ cell; I, interstitial or stromal tissue; L, lobule; M, mitotic figure; ME, gonadal mesentery / mesovary / mesorchium; O, oogonium; OL, ovarian lumen; PF, prefollicle cell; PO, pachytene oocyte; PS, pachytene spermatocyte; S, spermatogonium; SE, Sertoli cell; SO, somatic cell; SP, spermatocyst; SI, spermatocyte I; SII, spermatocyte II.*

For the first time, blood capillaries are observed in the gonadal dorsal region. The onset of ovarian meiosis is soon accompanied by an increase in germ cell mitoses. In addition, the ovary becomes either triangular or kidney-shaped in sections. A few days later, somatic reorganisation of the developing ovary is evidenced by somatic cell proliferation, which forms appendix-like structures which finally fuse to form the ovarian lumen.

The ovarian cavity is formed completely by day 65, when numerous basophilic oocytes at the diplotene meiotic stage are also present. From days 80 to 100, the ovary consists mainly of follicles which contain characteristically large, perinucleolar oocytes with a basophilic cytoplasm: these previtellogenic oocytes are arrested at the diplotene stage of the first meiotic prophase and are characterised by the presence of multiple round nucleoli in the peripheral nucleoplasm. Each perinucleolar oocyte is covered by a continuous layer of flat follicle cells. Interstitial connective tissue is present both in the ovarian periphery and in the angular spaces between three or more follicles in the inner ovarian region.

In contrast to gonadal development in the female, signs of histological differentiation in the prospective males are not observed until day 72. Therefore, undifferentiated gonads on day 65 are certainly diagnosed as prospective testes. They retain the pearlike shape of undifferentiated gonads and are much smaller than the ovaries of females of the same age. Spermatogonia at this stage cannot be distinguished from the undifferentiated germ cells at day 38, and their most characteristic feature is a prominent nucleus with one or two nucleoli.

By day 72, spermatogonial proliferation is occurring and blood vessels become evident in the dorsal region of the testis. On day 80, meiotic activity is apparent, and spermatocytes at early stages of meiotic prophase become numerous. Spermatocysts are formed by isogenic spermatocyte groups bound by the cytoplasmatic processes of Sertoli cells. Spermato-genesis progresses synchronously within each cyst. Spermatogonia are recognised as isolated cells, individually surrounded by the cytoplasm of a Sertoli cell. By day 100, the testis is divided into lobules composed by both Sertoli and germ cells, and which are separated by interstitial tissue. Spermatogenesis has reached the secondary spermatocyte stage.

Under laboratory conditions, mature gonads, i.e., ovaries containing full grown vitellogenic oocytes and testes containing mature spermatozoa can be observed in one year-old specimens, while the first spawnings are recorded towards month 16 after hatching.

The Adult Pituitary Gland

It is both an elongated and flat gland, in which the middorsal and caudal parts mostly consist of neurohypophyseal tissue. As it is found in other cichlids (e.g., Mattheij *et al.*, 1971) the infundibular recess of the third ventricle is large and extends as branches that penetrate the adenohypophyseal tissue.

The adenohypophysis is divided into the rostral pars distalis, the proximal pars distalis and the pars intermedia, the endocrine cells of these areas showing different tinctorial and immunohistochemical properties. The adenohypophysis is penetrated by neurohypo-physeal infoldings, particularly in the pars intermedia.

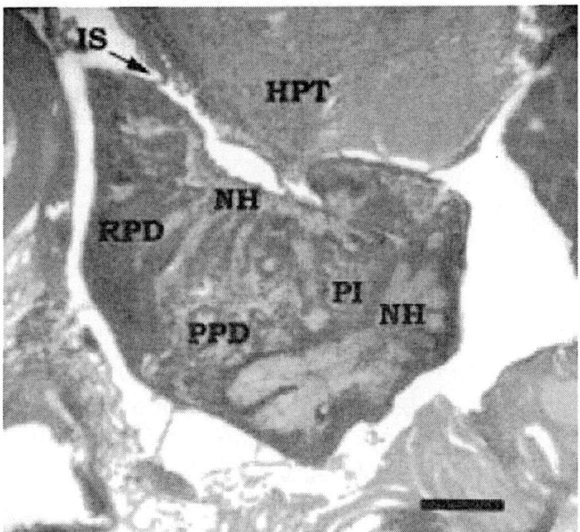

Figure: *Sagittal section of the pituitary gland of an adult male Cichlasoma dimerus. Hematoxylin-eosin. The adenohypophyseal lobes (rostral -RPD- and proximal pars distalis -PPD-, and pars intermedia-PI) are indicated. The infundibular stalk (IS) is shown connecting the hypothalamus (HPT) to the pituitary (arrow). The neurohypophysis (NH) also shows its numerous infoldings into the pars intermedia, and in the rostral and proximal pars distalis too. Scale bar = 50 μm.*

The distribution of endocrine cell types in the different adenohypophyseal lobes is summarised. Information about the antisera used in this study and the molecular weights of the detected *C. dimerus* pituitary hormones (Western blot analysis).

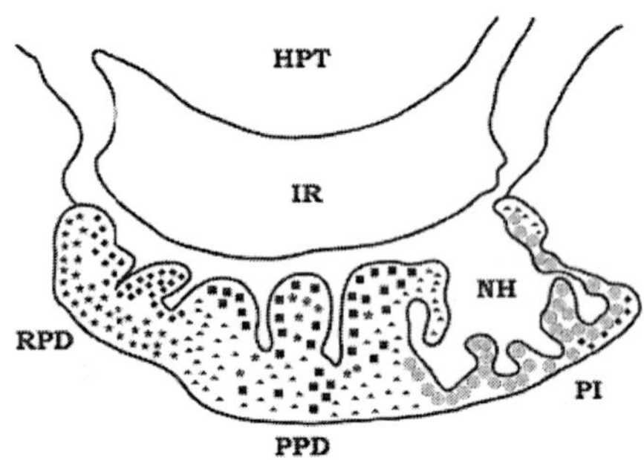

Figure: *Distribution map of hormone-secreting pituitary cells, as shown by their immunoreactivity to specific antisera. ACTH cells (Æ%) and PRL (&) cells occur in the rostral pars distalis. GH (*), TSH (%) and GTH(²%) cells are mainly located in the proximal pars distalis, while SL (Ï%) and a-MSH cells (Æ%) are located in the pars intermedia. HPT, hypothalamus; PI, pars Intermedia; RPD, rostral pars distalis; PPD, proximal pars distalis; NH, neurohypophysis; IR, infundibular recess. Redrawn from Pandolfi et al. (2001a).*

Rostral Pars Distalis

This lobe of the adenohypophysis is invaded by narrow neurohypophyseal prolongations (Pandolfi *et al.*, 2001a,b). In *C. dimerus,* two different cell types are typical of the rostral pars distalis, which are recognised by their immunoreactivity to antisera against prolactin and corticotropin (ir-PRL and ir-ACTH cells). Immunoreactive PRL cells are relatively large cells which form a compact layer in the periphery of the rostral pars distalis, while ir-ACTH cells are arranged in narrow bands or small follicles along the neurohypophyseal branches but without forming a continuous layer. Immunocytochemical labelling with the ACTH antibody was also observed in the pars intermedia, where the cells reacting with a-melanocyte stimulating hormone (ir-aMSH) cells are located. This cross-reactivity was likely due to the fact that both the corticotrops of the pars distalis and the melanotrops of the pars intermedia elaborate their distinctive hormones from the same large precursory molecule, namely propiomelanocortin (POMC) and that the whole a-MSH sequence is included in the ACTH mature molecule, so that the antisera against ACTH recognise a-MSH also (Eipper and Mains, 1981; Pandolfi *et al.*, 2003).

Overview of Our Studies in the Adult Pituitary Gland

A clear-cut lobar distribution of the common vertebrate pituitary hormones was found, with a localisation of ir-PRL and ir-ACTH cells in the rostral pars distalis, the glycoprotein hormones ir-FSH, ir-LH and ir-TSH cells located in the proximal pars distalis (together with ir-GH cells), while the ir-SL cells (which are specific of fish) are located in the pars intermedia, where they are together with the non fish-specific ir-aMSH cells.

In general, this lobar distribution would make the *C. dimerus* pituitary amenable to possibly interesting *in vitro* studies after microdissection and separate culture of the different lobes, since the chemically related PRL, GH and SL are produced in different lobes, as they are the chemically related ACTH and aMSH. Cells involved in the production of the glycoprotein family of gonadotropins and thyrotropin are located in a single lobe (the proximal pars distalis) but they would still be amenable to experimental manipulation with different putative releasing hormones.

Newly Hatched Larvae (1 h)

They show an incipiently differentiated brain, which is curved over the ventral part of the body, and a large yolk sac. At this stage, the brain and the pituitary gland cannot be distinguished from each other and no immunoreactive cells were detected.

Vitelline Larvae (1.5 days)

The brain becomes straightened out and its ventricles are highlighted by the actively proliferating cells which surround them. Near the prechordal cartilage, which is in contact with the diencephalon, a mass of presumptive adenohypophyseal cells appears, but none of them is reactive to PRL, GH or SL antisera. At this stage ir-aMSH and ir-ACTH cells are detected in the posterior and anterior part of the developing pituitary, respectively. No immunoreactivity to either b-FSH, b-LH or b-TSH was detected.

Vitelline Larvae (2 days)

The pituitary gland is recognised as a rounded structure hanging from the floor of the diencephalon and which is formed by morphologically similar epithelial cells. The neurohypophysis has not yet been differentiated. A few small ir-SL cells are scattered in the caudal part of the gland, showing strong immunoreactivity. Immunoreactive PRL cells also appear at this stage in the rostral area of the adenohypophysis. Some few ir-GH cells are immunostained weakly, and are distributed in the central area of the developing pituitary gland. At this stage, ir-aMSH and ir-ACTH cells are found in the posterior and anterior part of the developing pituitary, respectively.

Opened-Mouth Larvae (4 days)

The pituitary gland is located over a ventral cartilage plate, and it still shows no clearly delimited lobes. The detected pituitary cell types (ir-SL, ir-PRL and ir-GH) increase both in their number and their immunostaining intensity. A compact group of ir-PRL cells begins to form at the more rostral part of the gland, but these cells are not arranged in follicles as in adult animals. Both ir-SL and ir-GH cells begin to increase in number and size but still without forming compact groups. From this day on, a topographic segregation of the three cellular types in a rostral-caudal direction is observed, which precedes the adult lobar distribution.

Free-Swimming Larva (6 days)

The infundibular recess develops at the junction of the diencephalon and the pituitary gland. The adenohy-pophysis becomes larger but the neurohypophysis is not yet formed. The segregation of the endocrine cellular types (ir-ACTH, ir- aMSH, ir-GH, ir-PRL and ir-SL cells) is more evident at this stage, approximating their final localisations; also, however, some few immunore-active PRL, GH and SL cells remain isolated in different regions.

Odd-Fins Larva (14 days)

The neurohypophysis becomes distinct in the caudal part of the gland but it is still a thin structure and does not show the adult deep infoldings. The adenohy-pophysis becomes even larger and elongated in shape with ir-PRL and ir-ACTH cells occupying most of the rostral region (the prospective rostral pars distalis). The pars intermedia is distinct now, and is characterised by groups of ir-SL cells.

Prejuvenile Larva (21 days)

The first pituitary ir-FSH cells are detected in the anterior region of the adenohypophysis. From this stage on, the number of ir-FSH cells augments steadily. By day 90, when this ontogenetic study was ended, the number of FSH-ir cells was still increasing.

Prejuvenile Larva (26 days)

Two neurohypophyseal branches project into the adenohypophysis: a thicker one is directed towards the anterior adenohypophyseal regions while the other enters deep into the pars intermedia. At this time, a clear boundary between the pars distalis and the pars intermedia is seen, but there is still no clear demarcation between the rostral and the proximal pars distalis. Immunoreactive GH cells and SL-ir cells start to associate to the neurohypophyseal branches, both at the anterior adenohypophyseal regions and at the pars intermedia, respectively.

Juvenile (42 days)

The adult lobar topography is already distinct at this stage. Immunohistochemistry applied to adjacent sections showed that ir-SL cells are not periodate reactive, as they will be in the adult pituitary (Pandolfi *et al.*, 2001a). All the pituitary cell types, except ir-LH cells, are present at this stage. Close to the onset of sexual differentiation, which will occur by this time of development, the number of ir-FSH cells increases and they can be observed along the proximal pars distalis and the external border of the pars intermedia.

Juvenile (60 days)

Immunoreactive LH cells are first seen in central zones of the proximal pars distalis (two and a half weeks after the appearance of the first signs of ovarian differentiation).

Studies on Pituitary Ontogeny

Five immunoreactive cell types (PRL, GH, SL, ACTH, MSH) appear at very early stages of development, suggesting their functional importance for the fish larvae. Further studies focused on the ontogeny of hypothalamic factors that regulate their synthesis and secretion should still be performed. Gonadotrops become differentiated later: FSH cells appear prior to the onset of sexual differentiation, while LH cells appear during the sexual differentiation period.

Neuroendocrine Systems that Control Reproduction and Body Colour in C. Dimerus

Gonadotropin-Releasing Hormone (GnRH) Systems: The chemical structure of GnRH was initially discovered as a decapeptide which is present in the mammalian hypothalamus (Matsuo *et al.*, 1971; Burgus *et al.*, 1972) and is released in the portal system, and whose primary function was to induce pituitary gonadotropins release into the bloodstream. It was also found that it may act in neural tissue as a neuromodulator and/or neurotransmitter involved in eliciting reproductive behaviours (Ogawa *et al.*, 2006) and neuroendocrine reactions (Parhar, 1997). In addition to that, there is evidence suggesting that GnRH may play a role in the release of other pituitary hormones such as PRL, GH and SL (Weber *et al.*, 1997; Parhar, 1997; Stefano *et al.*, 1999; Vissio *et al.*, 1999).

Moreover, it was found that chemical forms other than the decapeptide also occurred in mammals and other vertebrates (Muske, 1993). At least three molecular GnRH forms exist in perciform species, which are identified after the species where they were originally found as sGnRH (salmon GnRH), sbGnRH (seabream GnRH) and cIIGnRH (chicken II GnRH). Their distribution patterns in the perciform central nervous system suggest the existence of three distinct neuronal populations located in (1) the olfactory bulbs, expressing sGnRH; (2) the preoptic area, predominantly expressing sbGnRH; and (3) the midbrain tegmentum, expressing cGnRH II.

Based upon these observations and those of ontogenetic studies, it has been hypothesized that sGnRH neurons originate from the olfactory placode, cGnRH II neurons originate from a mesencephalic primordium, and two possible origins were proposed for sbGnRH neurons: the diencephalic floor and the olfactory placode (Parhar, 1997; Parhar *et al.*, 1998; González-Martínez *et al.*, 2002). This hypothetical triple origin of GnRH neuronal systems differs from that proposed for some fish and other vertebrates, according to which all forebrain and diencephalic GnRH neurons would originate from the olfactory placode (Schwanzel-Fukuda and Pfaff, 1989; Wray *et al.*, 1989; Muske, 1993; Schwanzel-Fukuda, 1999).

In a recent study (Pandolfi *et al.*, 2005), both antibodies against different perciform GnRH-associated peptides (GAPs), and riboprobes to different perciform GnRH and GAPs were used. These GAPs are useful markers of the different GnRH neuronal populations since they are produced from GnRH precursor molecules and they are colocalised within GnRH-expressing cells (Ronchi *et al.*, 1992; Polkowska and Przekop, 1993). Furthermore, markedly different peptides are associated with the different GnRH types.

Therefore, these procedures result in a more specific marking than that obtained with antibodies and riboprobes to the smaller GnRH molecules themselves, both for immunohistochemistry and for *in situ* hybridization, respectively. Results showed that distribution of both sGnRH and sbGnRH neuronal populations in the *C. dimerus* forebrain are largely overlapped along the olfactory bulb, the nucleus olfacto-retinalis, the preoptic area and the ventral telencephalon. It was also shown (Pandolfi *et al.*, 2005) that sGAP immunoreactivity follows an antero-posterior gradient, with the most conspicuous immunoreactivity (both in terms of cell area and number) in the nucleus olfactoretinalis, and a lesser one in more posterior locations, as the preoptic area and the ventral telencephalon. In contrast, sbGAP cells show a postero-anterior gradient, with the most conspicuous immunoreactivity in the preoptic area and a weaker one in the nucleus olfactoretinalis and olfactory bulb.

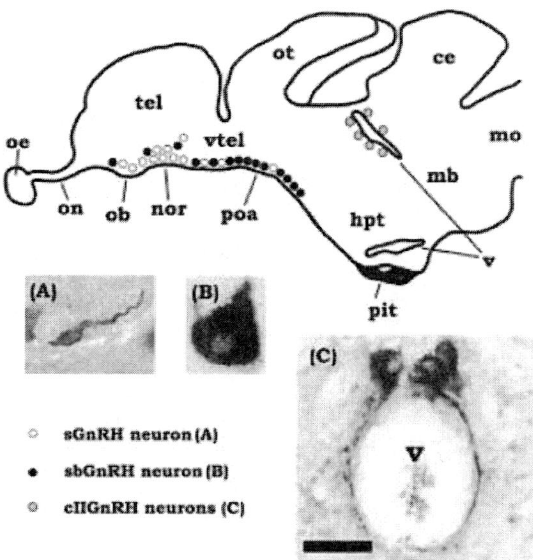

Figure: *Distribution map of neurons producing the different types of GnRH found in the brain of Cichlasoma dimerus. The location of sGnRH, sbGnRH and cIIGnRH neurons was studied by immunoreactivity of their respective GnRH-associated peptides (GAPs). Insets A, B and C show details of sGnRH, sbGnRH, and cIIGnRH neurons at the same magnification; scale bar = 20 μm. Encephalic regions and brain nuclei are indicated as: ce, cerebellum; hpt, hypothalamus; mo, medulla oblongata; mb, midbrain; nor, nucleus olfacto retinalis; ob, olfactory bulb; oe, olfactory epithelium; on, olfactory nerve; ot, optic tectum; pit, pituitary; poa, preoptic area, tel, telencephalon; v, ventricle; vtel, ventral telencephalon. Redrawn from Pandolfi et al. (2005).*

Overview of Our Studies on Neuroendocrine Systems that Control Reproduction (GnRH)

Our results showed that: (1) GnRH neuronal populations in the forebrain (sGAP and sbGAP) show an overlapping pattern, (2) sGAP projections are mainly located in the forebrain and contribute to the pituitary innervation, with cGAPII projections being mainly distributed along the mid and hindbrain and not contributing to pituitary innervation, whereas sbGAP projections are restricted to the ventral forebrain, being the most important molecular form in relation to pituitary innervation; (3) sbGAP neurons have an olfactory origin; (4) GAP antibodies and GAP riboprobes are valuable tools for the study of various GnRH systems, by avoiding the cross-reactivity problems that occur when using GnRH antibodies and GnRH riboprobes alone.

Regulation of SL Release by GnRH and MCH

As we have seen, many studies have focused on SL function, but there is scarce information on the regulation of its secretion. In cultured pituitaries of the rainbow trout, *Oncorhynchus mykiss*, Kakizawa *et al.* (1997) have shown that sGnRH, corticotropin releasing factor and serotonin stimulate SL release. Also, in the masu salmon, *Oncorhynchus masou*, Onuma *et al.* (2005) observed that the addition of sGnRH to primary pituitary cell cultures elevated SL mRNA expression.

In *C. dimerus*, we have shown by double label immunofluorescence that ir-SL cells are in close contact with both ir-MCH and ir-GnRH fibres in the pars intermedia (Cánepa *et al.*, 2008). We then performed *in vitro* pituitary cultures of intact in order to investigate a possible releasing or inhibiting function of MCH or GnRH on SL cells. We found that MCH stimulated SL release from pituitaries of male *C. dimerus* in a dose dependent manner; but no such an effect was observed in females. On the other hand, incubation with GnRH showed similar results than the MCH assay in male pituitaries. Although female pituitaries did not evidence significant differences between treatments, we observed a noticeable increasing tendency. In conclusion, our results show that SL, like other pituitary hormones, would be under hypothalamic control (MCH and GnRH) and that they are implicated in diverse physiological processes including background adaptation and reproduction (Cánepa *et al.*, 2008).

The Roles of Pituitary and Brain Gonadotropins.

The gonadotropins follicle-stimulating hormone (FSH) and luteinizing hormone (LH) are glycoprotein hormones produced in the proximal pars distalis of *C.dimerus*. In fishes, as in other vertebrates, these hormones are considered of central importance in the control of gametogenesis and gonadal steroidogenesis (Blázquez *et al.*, 1998). In several tetrapod and teleost species, they are formed by two protein chains, namely the a and b subunits (Yoshiura *et al.*, 1999; Yaron *et al.*, 2001). The a-FSH and a-LH subunits have identical aminoacid sequences in a given species, while b subunits have different sequences, and are recognised by different specific receptors, and are thus responsible for their different actions.

In salmonid species characterised for spawning only once a year, the different roles of FSH and LH have been studied in detail (Suzuki *et al.*, 1988; Tyler *et al.*, 1991). The functions of FSH, as well as the variations of its circulating levels during the reproductive cycle, are not completely understood in many non salmonid species examined so far, though FSH is generally considered as a vitellogenic hormone, while LH is generally related to gonadal maturation and steroidogenesis.

Several developmental studies conducted in teleost fishes, showed varying plasma levels and different expression patterns of FSH and LH in the pituitary (Dufour *et al.*, 2000). Taken together, these results suggest specific functions for both FSH and LH at the onset of sexual maturation and in the control of gametogenesis. As we have mentioned above, the identification of FSH cells using heterologous antisera was very difficult to perform due to cross-reactions or low specificity between FSH molecules from different species (Vissio *et al.*, 1996; Pandolfi *et al.*, 2001a). Antisera raised against the mummichog *Fundulus heteroclitus* FSH and LH were successfully used to recognise GtH cells in several teleost fishes, because they recognised conserved regions of FSH and LH in teleosts.

The localisation, characterisation and ontogeny of gonadotrops in the pituitary of *Cichlasoma dimerus* have been already addressed this review. In this species, pituitary ir-FSH cells appear before the onset of sexual differentiation, while ir-LH cells differ-entiate several days thereafter. This variation in the timing of immunoreactive FSH and LH expression was also observed in several teleost species (Mal *et al.*, 1989; Nozaki *et al.*, 1990; Saga *et al.*, 1999; Magliulo-Cepriano *et al.*, 1994). The results of these studies suggested that in salmonid species and in the platyfish (Schreibman *et al.*, 1982; Parhar *et al.* 1995), as well as in *C. dimerus*, the differentiation of ir-FSH cells precedes that of ir-LH cells. In the pejerrey *Odontesthes bonariensis*, however, ir-LH cells are known to appear before ir-FSH cells.

In salmonids, FSH, but not LH, is involved in the onset of gonadal growth and development (Saga *et al.*, 1993). Also, FSH was shown to be involved in the onset of meiosis in an amphibian (Ito and Abe, 1999). Since ir-FSH was expressed in pituitary cells of sexually undifferentiated prejuvenile larvae of *C. dimerus* (30 days after hatching), we decided to study the effects of this hormone on larval gonadal explants. LH effects was not studied since ir-LH cells do not appear in the pituitary until day 60, i.e. two and a half weeks after the first indications of gonadal differentiation appear. When the undifferentiated gonadal explants were left for 15 days with no hormone in the control culture medium, no sexual differentiation occurred, while exposure to human recombinant FSH for the same period resulted in 80% of cases differentiating as prospective ovaries (N=80) (Pandolfi *et al.*, 2006). These results reinforced our previous hypothesis about a temporal correlation between the innervation of pituitary cells by sbGnRH fibres and the subsequent release of FSH to circulation, and the onset of gonadal sex differentiation in both sexes.

In the course of our anatomical study of the *C. dimerus* pituitary (Pandolfi *et al.*, 2001a; 2006), ir-FSH and ir-LH fibres were detected in both the neurohypo-physis and the infundibular recess. These fibres were deriving from cell bodies in the parvicelullar and magnocellular preoptic nuclei, and in the hypothalamic nucleus lateralis tuberis. Also, the developmental appearance of both brain- and pituitary-derived gonadot-ropins was studied showing that ir-LH neurons appeared on day 15 after hatching, while ir-FSH pituitary cells and neurons appeared on day 21), and ir-LH pituitary cells appeared only on day 60. To corroborate these findings, specific PCR products of b-LH and b-FSH were amplified from cDNA obtained from pituitaries and the preoptic area and hypothalamus of adult *C. dimerus* during the reproductive season. The partial nucleotide sequences of mRNA from *C. dimerus* brain and from the preoptic-hypothalamic brain fragment consisted of approximately 258 bp for b-LH and 276 bp for b-FSH (Pandolfi *et al.*, 2009).

Furthermore, we settled an individual pituitary *in vitro* culture system to study the possible modulatory effect of brain derived gonadotropins on pituitary hormone secretion. Whole pituitary explants were cultured with different concentrations of *Fundulus heteroclitus* LH or FSH and the culture media were analyzed by Western blot. Exogenous LH produced a dose-dependent increase in pituitary b-LH and SL release and an increase in b-FSH release. No effect was observed on both GH and PRL release. Exogenous FSH produced an inhibition in b-LH release, a dose-dependent increase in b-FSH and SL release, and no effect on PRL and GH release. This is the first work showing an effect of brain-derived gonadotropins on different vertebrate pituitary cells (Pandolfi *et al.*, 2009).

Attempt to Improve the Reproductive Efficiency of Nile Tilapia Brood Stock Fish

Tilapia aquaculture is and will continue to be an important fish, particularly for the lesser-developed countries in the tropics. Nile tilapia (*Oreochromis niloticus)* are considered as the most common and popular fish in Egypt. Egypt, a country where, arguably, the farming of tilapia has its roots, where tilapia culture is believed to have originated some 4000 years ago. Tilapia consist 36% of the Egyptian production from fish culture and occupy the 10th order concerning the world production from aquaculture. Hence, Egypt produces 20% of the world tilapia capture and 12% of the world farmed tilapia.

The culture of *O. niloticus* in Egypt has recently developed into a major industry. This industry, however, is still growing in a remarkable way with apparent intend towards intensification that pressurizing the need of enormous number of seeds. Many limitations associated tilapia fry production under the prevailing Egyptian conditions were described by El-Gamal. Also, brood stock husbandry and spawning techniques are constantly upgraded as Egyptian hatcheries require a high number of good quality eggs to satisfy the needs for aquaculture, so rigorous management of large numbers of brood stock are necessary for mass production of eggs and fry due to the complex reproductive biology and asynchronously spawning with relatively small number of eggs produced per spawning.

Accordingly, the development of more elaborated forms of brood stock management is crucial to improve fry yield and system efficiency. Today, it is widely accepted that effective seed production demands a thorough understanding of the special husbandry and particular nutritional requirements of brood stock fish which significantly affect fecundity, survival, egg size and egg and larval quality. The objective of the present research was to study the possibility of improving reproductive performance of Nile tilapia fish using some feed additives.

6

Cell Biology of Oxytocin and Vasopressin Cells in Fishes

The neurohypophysial peptide oxytocin (OT) and OT-like hormones facilitate reproduction in all vertebrates at several levels. The major site of OT gene expression is the magnocellular neurons of the hypothalamic paraventricular and supraoptic nuclei. In response to a variety of stimuli such as suckling, parturition, or certain kinds of stress, the processed OT peptide is released from the posterior pituitary into the systemic circulation. Such stimuli also lead to an intranuclear release of OT. Moreover, oxytocinergic neurons display widespread projections throughout the central nervous system. However, OT is also synthesized in peripheral tissues, e.g., uterus, placenta, amnion, corpus luteum, testis, and heart. The OT receptor is a typical class I G protein-coupled receptor that is primarily coupled via G_q proteins to phospholipase C-β.

The high-affinity receptor state requires both Mg^{2+} and cholesterol, which probably function as allosteric modulators. The agonist-binding region of the receptor has been characterised by mutagenesis and molecular modelling and is different from the antagonist binding site. The function and physiological regulation of the OT system is strongly steroid dependent. However, this is, unexpectedly, only partially reflected by the promoter sequences in the OT receptor gene. The classical actions of OT are stimulation of uterine smooth muscle contraction during labour and milk ejection during lactation. While the essential role of OT for the milk let-down reflex has been confirmed in OT-deficient mice, OT's role in parturition is obviously more complex. Before the onset of labour, uterine sensitivity to OT markedly increases concomitant with a strong upregulation of OT receptors in the myometrium and, to a lesser extent, in the decidua where OT stimulates the release of $PGF_{2\alpha}$. Experiments with transgenic mice suggest that OT acts as a luteotrophic hormone opposing the luteolytic action of $PGF_{2\alpha}$. Thus, to initiate labour, it might be essential to generate sufficient $PGF_{2\alpha}$ to overcome the luteotrophic action of OT in late gestation. OT also plays an important role in many other reproduction-related functions, such as control of the estrous cycle length, follicle luteinization in the ovary, and ovarian steroidogenesis. In the male, OT is a potent stimulator of spontaneous erections in rats and is involved in ejaculation. OT receptors have also been identified in other tissues, including the kidney, heart, thymus, pancreas, and adipocytes. For example, in the rat, OT is a cardiovascular hormone acting in concert with atrial natriuretic peptide to induce natriuresis and kaliuresis. The central actions of OT range from the modulation of the neuroendocrine reflexes to the establishment of complex social and bonding behaviours related to the reproduction and care of the offspring. OT exerts potent antistress effects that may facilitate pair bonds. Overall, the regulation by gonadal and adrenal steroids is one of the most remarkable features of the OT system and is, unfortunately, the least understood. One has to conclude that the physiological regulation of the OT system will remain puzzling as long as the molecular mechanisms of genomic and nongenomic actions of steroids have not been clarified.

The neurohypophysial hormone oxytocin (OT) was the first peptide hormone to have its structure determined and the first to be chemically synthesized in biologically active form. It is named after the "quick birth" which it causes due to its uterotonic activity. OT was also found to be responsible for the milk-ejecting activity of the posterior pituitary gland. The structure of the OT gene was elucidated in 1984, and the sequence of the OT receptor was reported in 1992.

OT is a very abundant neuropeptide. This became obvious in a study where the most prevalent hypothalamic-specific mRNAs were analysed. OT was found to be the most abundant of 43 transcripts identified. Today, we recognise that OT exerts a wide spectrum of central and peripheral effects. The actions of OT range from the modulation of neuroendocrine reflexes to the establishment of complex social and bonding behaviours related to the reproduction and care of the offspring. Overall, the cyclic nonapeptide OT and its structurally related peptides facilitate the reproduction in all vertebrates at several levels.

In the following two sections, we describe the structural features of the OT receptor system on the molecular level. OT has long been considered to be restricted to stimulation of uterine contractions during labour and milk ejection during lactation. These classical functions of the OT system are treated in the first parts of section IV. The fact that OT is found in equivalent concentrations in the neurohypophysis and plasma of both sexes suggests that OT has further physiological functions. The expression of OT and its receptor has now been identified in a variety of peripheral tissues.

For example, evidence was provided that OT acts in concert with atrial natriuretic peptide (ANP) in the control of body fluid and in cardiovascular homeostasis in the rat. It is important to note that most of our knowledge about functions of OT derives from studies with rats. However, localisation and expression patterns of OT receptors show marked species differences, suggesting that some of the described OT activities may be species specific. Over the past decade, particularly the central actions of OT have been intensively studied revealing a profound regulation by steroids. The regulation by steroids is in fact a common theme throughout the review and is probably the most remarkable feature of the OT receptor system. Finally, in the last section, we summarise the contributions that have been reported on behavioural effects mediated or modulated by OT. Ironically, the classical "oxytocic" function of OT is again open for discussion due to the results with OT-deficient mice.

Oxytocin and Oxytocin-like Peptides

Evolutionary Aspects A

All neurohypophysial hormones are nonapeptides with a disulfide bridge between Cys residues 1 and 6. This results in a peptide constituted of a six-amino acid cyclic part and a COOH-terminal α-amidated three-residue tail. Based on the amino acid at position 8, these peptides are classified into vasopressin and OT families: the vasopressin family contains a basic amino acid (Lys, Arg), and the OT family contains a neutral amino acid at this position. Isoleucine in position 3 is essential for stimulating OT receptors and Arg or Lys in position 8 for acting on vasopressin receptors. The difference in the polarity of these amino acid residues is believed to enable the vasopressin and OT peptides to interact with the respective receptors.

Virtually all vertebrate species possess an OT-like and a vasopressin-like peptide. Bony fishes (Osteichthyes), predecessors of the land vertebrates, possess isotocin and vasotocin. Thus two evolutionary molecular lineages have been proposed: an isotocin-mesotocin-OT line, associated with reproductive functions, and a vasotocin-vasopressin line involved in water homeostasis. Because vasotocin has been found in the most primitive cyclostomes, the OT and vasopressin genes may have arose by duplication of a common ancestral gene after the radiation of cyclostomes. Based on calculations from the nucleotide level of OT and vasopressin gene, the ancestral gene encoding the precursor protein should be more than 500 million years old.

The exceptional structural stability of the nonapeptides during evolution suggests a strong selective pressure, e.g., by coevolution with the corresponding receptors and/or with specific processing enzymes. A conspicuous diversity of OT-like peptides is found in cartilaginous fishes (Chondrichthyes). These marine fishes use urea rather than salts for osmoregulation. It was hypothesized that the OT-like hormones gained their high diversity in Chondrichthyes as they have been relieved from the control of ionic homeostasis. Notably, OT, the typical hormone of placental mammals, has been identified in the Pacific ratfish, a Chondrichthyes species.

Mesotocin is the OT-like hormone found in most terrestrial vertebrates from lungfishes to marsupials, which includes all nonmammalian tetrapods (amphibians, reptiles, and birds). Only two South American marsupials express OT exclusively, whereas all other marsupials have mesotocin. In the Northern brown bandicoot (*Isoodon macrourus*) and the North American opossum (*Didelphis virginiana*), OT is present together with mesotocin. Overall, mesotocin has the largest distribution in vertebrates after vasotocin found in all nonmammalian vertebrates and isotocin identified in bony fishes. Despite this invariability, no clear physiological role has been ascribed to this peptide so far. It is unknown whether the marsupial species that are endowed with both OT and mesotocin have two distinct receptors. The earthworm *Eisenia foetida* is the most primitive species from which an OT-related peptide (annetocin) has been isolated. Injection of annetocin in the earthworm or in leechs results in induction of egg-laying behaviour.

Gene Structure B

In all species, OT and vasopressin genes are on the same chromosomal locus but are transcribed in opposite directions . The intergenic distance between these genes range from 3 to 12 kb in mouse, human, and rat. This type of genomic arrangement could result from the duplication of a common ancestral gene, which was followed by inversion of one of the genes. The human gene for OT-neurophysin I encoding the OT prepropeptide is mapped to chromosome 20p13 and consists of three exons: the first exon encodes a translocator signal, the nonapeptide hormone, the tripeptide processing signal (GKR), and the first nine residues of neurophysin; the second exon encodes the central part of neurophysin (residues 10–76); and the third exon encodes the COOH-terminal region of neurophysin (residues 77–93/95).

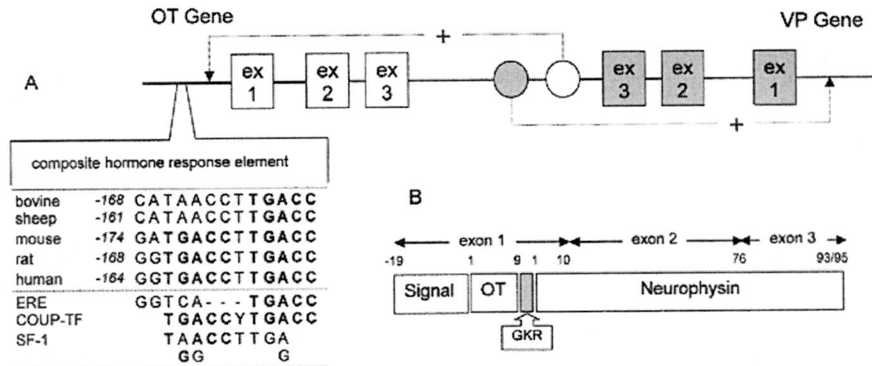

Figure: Organisation of the oxytocin (OT) and vasopressin (VP) gene structure including schematic depiction of the putative cell-specific enhancers (open circle, enhancer of OT gene; shaded circle, enhancer of VP gene). A: *details of the approximately "160-bp region (composite hormone response element) of the upstream OT gene promoter conserved across five species including the sequences of the response elements estrogen response element (ERE), chicken ovalbumin upstream promoter transcription factor I (COUP-TF), and steroidogenic factor-1 (SF-1) are indicated.* B: *domain organisation of preprooxytocin including the processing sites. The precursor is split into the indicated fragments by enzymatic cleavages, one involving a glycyl-lysyl-arginine (GKR) sequence and leaving a carboxamide group at the COOH-terminal end of OT. Signal, signal peptide.*

The high homology of the OT-like precursor polypeptides is well documented in the sequence of preproannetocin from *Eisenia foetida*, a primitive invertebrate. It consists of a signal peptide, annetocin (flanked by a Gly COOH-terminal amidation signal and a Lys-Arg dibasic endoproteolytic sequence), and a neurophysin domain. Notably, 14 cysteine residues that play a crucial role in constructing the correct tertiary structure of a neurophysin are completely conserved in the *Eisenia* neurophysin domain.

The OT prepropeptide is subject to cleavage and other modifications as it is transported down the axon to terminals located in the posterior pituitary. The mature peptide products, OT and its carrier molecule neurophysin, are stored in the axon terminals until neural inputs elicit their release. The main function of neurophysin, a small (93–95 residues) disulfide-rich protein, appears to be related to the proper targeting, packaging, and storage of OT within the granula before release into the bloodstream. OT is found in high concentrations (>0.1 M) in the neurosecretory granules of the posterior pituitary complexed in a 1:1 ratio with neurophysin. In such complexes, OT-neurophysin dimmers are the basic functional units as suggested by the crystal structure of the neurophysin-OT complex. Cys-1 and Tyr-2 in the OT molecule are the principal neurophysin binding residues. In particular, the protonated α-amino group (Cys-1) in OT forms an essential contact site to neurophysin via electrostatic and multiple hydrogen bonding interactions. Due to its dependence on amino group protonation (pK_a ~6.4), the binding strength between OT and neurophysin is much higher in an acidic compartment like the neurosecretory granules (pH ~5.5). Conversely, the dissociation of the complex is facilitated as the complex is released from the neurosecretory granules and enters the plasma (pH 7.4).

Gene Regulation C

Due to the lack of an appropriate cell culture system, the regulation of OT gene expression was studied in heterologous systems and in transgenic mice. The expression patterns found with bovine OT transgenes in mice suggested very complex mechanisms for the cell type-specific expression of OT genes. A bovine OT transgene consisting of the OT structural gene flanked by 600 bp of upstream and 1,900 bp of downstream sequences contained sufficient information to direct expression to murine oxytocinergic magnocellular neurons, within which it was subject to physiological regulation. However, enlargement of OT transgene constructs by addition of 700 bp of contiguous downstream sequences repressed the hypothalamic expression. Analysis of various gene constructs in transgenic mice led to the proposal that cell-specific enhancers for OT and vasopressin gene expression are not located on the 52 -upstream regions of these genes, but are present in the intergenic region 0.5–3 kb downstream of the vasopressin gene. So, constructs containing genomic DNA from 0.5 to 9 kb 52 -upstream of the OT and vasopressin genes but with no endogeneous 32 -downstream sequences did not show significant expression in the hypothalamic magnocellular neurons.

The OT mRNA in the rat shows an increase in poly (A) tail length in response to the activation of the hypothalamoneurohypophysial system, e.g., during pregnancy, lactation, and dehydration. This could augment mRNA stability and may be an additional level of OT gene control. Hexanucleotide AGGTCA motifs and variations thereof are present in the proximal 52 -flanking region of cloned OT genes. This motif is part of binding sites for all members of the nuclear receptor superfamily, except the glucocorticoid, mineralocorticoid, progesterone, and androgen receptor.

Various combinations of this motif exist, ranging from single hexanucleotides, direct or inverted repeats with spacing varying from one to at least six nucleotides. Thus potentially several members of the nuclear receptor family including many orphan receptors could interact with the OT gene and regulate its expression. The human and rat OT promoters could be stimulated by the ligand-activated estrogen receptors ERα and ERβ, the thyroid hormone receptor THRα, and the retinoic acid receptors RARα and RARβ in a variety of cells. However, it is important to note that these results were obtained from cotransfection experiments in cell lines, i.e., under nonphysiological circumstances.

A highly conserved DNA element exists at -160 nucleotides upstream from the transcriptional initiation site. Deleting the region between 172 and 148 resulted in complete loss of thyroid hormone responsiveness and most of the responsiveness to estrogen and retinoic acid. This special "composite" hormone response element is composed of three TGACC motifs. Two of them form an inverted repeat with a spacing of three nucleotides that differs in one nucleotide from the palindromic, canonical estrogen response element (ERE). The rat and human OT promoter shows a good homology with the classic palindromic ERE. Accordingly, in a heterologous transfection system, the rat and human but not the bovine OT gene promoter could be stimulated by estradiol. In the appropriate cellular context, the strongest activators were ERα and ERβ. The composite hormone response element was suggested to synergise with proximal elements for the estrogen responsiveness and was found to be essential for the positive regulation by retinoic acid. However, estrogen receptor expression was not detected in oxytocinergic cells of the rat hypothalamus. Thus direct estrogen-dependent activation may not regulate the OT gene expression in magnocellular neurons in vivo. An estrogen responsiveness was reported for a parvocellular OT-expressing cell group that contains ERβ. Although THRs have been localised in supraoptic nucleus (SON) and paraventricular nucleus (PVN), only a small influence of thyroid hormones on OT gene expression was found in rats in vivo.

The rat uterus displays a marked upregulation of OT gene expression before delivery. The main site of steroid-induced uterine OT gene expression was the endometrial epithelium. A strong increase in OT mRNA (150 fold) preceded the increase in uterine OT binding sites that occurs very shortly before the onset of labour. The estrogen-induced rise in uterine OT mRNA was probably mediated via the common hormone response element in the OT gene promoter. The palindromic structure at the composite hormone response element at approximately –160 bp was identified as necessary and sufficient for estrogen induction of the OT gene promoter. However, the high level of uterine OT mRNA at term was not achieved by any of the steroid treatment regimens tested so far.

Several investigations have focused on the role of nuclear orphan receptors in the regulation of the OT gene. Unlike in the hypothalamus, the bovine OT gene could be tissue-specifically expressed in the gonads by a minimal functional promoter contained within 600 bp of the transcription start site. As mentioned above, the bovine OT gene is unresponsive to estradiol. Nuclear orphan receptors of corpus luteum granule cells have been identified to interact with the common hormone response element: chicken ovalbumin upstream promoter transcription factor I (COUP-TFI) and steroidogenic factor-1 (SF-1). The levels of these factors could be responsible for the regulation of the endogeneous OT gene in this tissue. SF-1 is a factor with constitutive activating properties on the OT gene. The orphan COUP-TFI repressed the activation of the rat OT gene induced by retinoic acid, thyroid hormone, and estrogens through competitive binding to the composite hormone response element. Another orphan receptor identified in the hypothalamus, testis receptor 4 (TR4), interacts, unlike all other nuclear receptors, at a region further downstream (–112/"77 bp) of the common response element in the OT gene. The 52 -flanked region of the rat OT gene also contains binding sites for class III POU homeodomain proteins. Brn-2, a member of this family, is involved in the regulation of OT genes in magnocellular neurons. Brn-2 null mice lack magnocellular vasopressin- and OT-expressing neurons in SON and PVN. Brn-2 is required for a specific step in the developmental fate of magnocellular neurons. However, POU class III proteins did not display a significant regulatory activity on the OT gene in heterologous expression systems. Taken together, OT gene regulation in vivo appears to be governed by multiple enhancers and repressors interacting in a complex yet ill-defined fashion.

Oxytocin Receptors

Gene Structure and Regulation A

Kimura et al. first isolated and identified a cDNA encoding the human OT receptor using an expression cloning strategy. The encoded receptor is a 389-amino acid polypeptide with 7 transmembrane domains and belongs to the class I G protein-coupled receptor (GPCR) family. To date, the OT receptor encoding sequences from pig, rat, sheep, bovine, mouse, and rhesus monkey have also been identified.

The human OT receptor mRNAs were found to be of two sizes, 3.6 kb in breast and 4.4 kb in ovary, endometrium, and myometrium. The OT receptor gene is present in single copy in the human genome and was mapped to the gene locus 3p25-3p26.2. The gene spans 17 kb and contains 3 introns and 4 exons. Exons 1 and 2 correspond to the 52 -prime noncoding region. Exons 3 and 4 encode the amino acids of the OT receptor. Intron 3, which is the largest at 12 kb, separates the coding region immediately after the putative transmembrane domain 6.

Exon 4 contains the sequence encoding the seventh transmembrane domain, the COOH terminus, and the entire 32 noncoding region, inluding the polyadenylation signals. Although many GPCRs have an intronless gene structure, the genes for some other members of the GPCR family including the human vasopressin V2 receptor contain an intron at the same location after transmembrane domain 6. The transcription start sites lie 618 and 621 bp upstream of the initiation codon as demonstrated by primer extension analysis. Nearby, a TATA-like motif and a potential SP-1 binding site is found in the human OT receptor gene. The 52 -flanking region also contains invert GATA-1 motifs, one c-Myb binding site, one AP-2 site, two AP-1 sites, but no complete ERE. Instead, there were two half-palindromic 52 -GGTCA-32 motifs and one half-palindromic 52 -TGACC-32 motif of ERE. Moreover, there were two nucleofactor interleukin-6 (NF IL-6) binding consensus sequences and two binding site sequences for an acute phase reactant-responsive element at the 52 -flanking region.

In the OT receptor gene of the mouse, the promoter region lacks an apparent TATA box but contains multiple putative interleukin-response elements, several half-palindromic motifs, and a classical ERE.

In case of the rat OT receptor gene expression at parturition, three transcripts (2.9, 4.8, and 6.7 kb) were identified that differ in the length of their 32 -untranslated regions. The promoter region of the rat OT receptor gene also contains multiple putative interleukin-response elements, NF IL-6, and acute-phase response elements (APRE). Further sequence analysis of 4 kb of the 52 -flanking DNA of the rat OT receptor gene revealed the presence of a cAMP response element (CRE) as well as several other potential regulatory elements, including AP-1, AP-2, AP-3, AP-4 sites, an ERE, and a half-steroid response element. A palindromic ERE was identified -4 kb 52 of the translational start site. An OT receptor reporter construct of this promoter in the human cancer breast cell line MCF-7 demonstrated pronounced induction by both forskolin and phorbol ester, but contrary to in vivo findings, only a weak transcriptional response to estradiol.

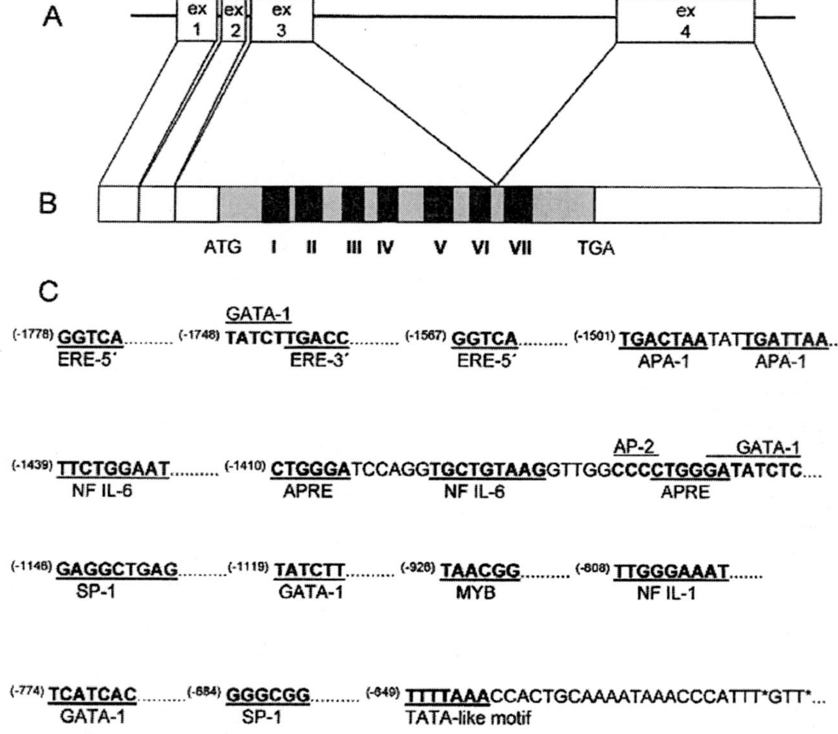

Figure: *Organisation of the human OT receptor gene including the localisation of consensus sequences for transcription factors. The human OT receptor gene consists of four exons. Exons 3 and 4 encode the amino acid sequence for the OT receptor. The start (ATG) and stop (TGA) codons of the receptor cDNA are indicated. The DNA sequences encoding for transmembrane regions I–VII are indicated by black areas.*

Constructs of the CRE and half-steroid response elements from the promoter of the rat OT receptor gene function as active enhancers. This suggests a potential role for protein kinase A and C pathways in OT receptor gene regulation. The protein kinase A pathway, for example, may be activated when forskolin treatment promotes upregulation of OT receptors in cultured rabbit amnion cells. Protein kinase C may act to increase fos/jun activity at AP-1 sites in response to phorbol ester treatment. In a mammary tumor cell line (Hs578T) that expresses inducible, endogenous OT receptors, a DNA region containing an ets family target sequence, and a CRE/AP-1-like motif was required for both basal and serum-induced OT receptor gene expression. The ets factor GABPα/β slightly induced OT receptor gene expression in this human breast cell line. The gene expression was markedly potentiated following cotransfection with c-*fos*/c-*jun*.

APREs are typically found in genes for acute-phase proteins such as α_2-macroglobulin or T kininogen, which are induced by infection or inflammation. The presence of these elements in the promoter region of the human and rat OT receptor gene suggests that the acute induction of OT receptor expression could be a phenomenon similar to the induction of acute-phase reponse genes. The decidua has "macrophage-like" properties and functions. Possibly, inflammatory cytokines are able to induce labour and, thereby, take usage of the transcriptional activation of the OT receptor gene.

However, the lack of classical EREs in a promoter does not exclude a potential direct effect of estrogens on gene expression, since the present half-palindromic ERE motifs can also act synergistically to mediate estrogen activation as shown in the ovalbumin gene. Gonadal steroids have an important influence on the uterine OT receptor mRNA accumulation in vivo. Estrogens administered to ovariectomised rats increased OT receptor binding sites and increased OT receptor mRNA accumulation severalfold. Although progesterone leads to a marked decline of OT receptor binding sites, the mRNA levels of OT receptor were nearly unchanged. This and several other findings imply the involvement of nongenomic effects of progesterone.

The OT receptor gene is differentially expressed in various tissues. In uterus or hypothalamus, the OT receptor regulation correlates with the pattern of sex steroids, in particular estradiol. As shown with knock-out mice, ERα is not necessary for basal OT receptor synthesis but is absolutely necessary for the induction of OT receptor binding in the brain by estrogen. However, it is unclear whether OT receptor gene transcription is predominantly regulated by estrogen. The continuous presence of receptors in certain brain regions after gonadectomy suggests the existence of alternate mechanisms of regulation. In this context, a study of the tammar wallaby, an Australian marsupial, is interesting.

This species has a twin uterus attached to a double cervix so that each uterus forms an independent environment. During pregnancy, only one uterus becomes gravid, the other remains empty and can thus be regarded as a natural control for uterine changes occurring during pregnancy. It was shown that mesotocin receptor concentrations and the responsiveness to mesotocin differed between the gravid and nongravid myometrium during pregnancy. This indicates that the stimulating agent for the mesotocin receptor is unlikely a circulating factor but rather a local factor, possibly of fetal or placental origin. Another well-studied system is the OT receptor gene expression in bovine endometrial cells. In vivo, bovine endometrial OT receptors are upregulated in a cycle-dependent fashion. This regulation appears to be completely at the transcriptional level. Even if the receptors are downregulated in vivo, they show upregulation when explanted and cultured in vitro.

This indicates that the OT receptor regulation is partly due to gene suppression in vivo. Despite the presence of steroid receptors in bovine endometrial cells, the level of OT receptor mRNA could neither be affected by progesterone or estradiol nor by a progesterone withdrawal protocol. The only factor that affected the OT receptor mRNA level was interferon-[3]. As in vivo, this cytokine suppressed the OT receptor mRNA production. Nuclear protein binding and transfection experiments suggested that constitutive upregulation is a feature of the OT receptor promoter.

So, specific gene suppression is likely to play an important role for physiological control of the OT receptor expression. Of interest, a genomic element within the third intron of the human OT receptor gene was found to be associated with transcriptional gene suppression. The intronic region was hypermethylated in nonexpressing tissues, but relatively hypomethylated in the myometrium of the cycle and at term, when the OT receptor gene is upregulated. Taken together, it was concluded that sex steroids have an indirect effect on both the OT and OT receptor genes, possibly involving intermediate transcription factors or cofactors. The transcriptional regulation of OT receptor shows species-specific differences. The brain OT receptor varies across species in its distribution as well as in its regional regulation by gonadal steroids. For the OT receptor as for many other genes, the DNA sequences located in the 52 -flanking region upstream from the coding region are primarily responsible for conferring tissue-specific expression. Transgenic mice carrying 5 kb of the 52 -flanking region of the prairie vole OT receptor gene showed the typical expression pattern of prairie vole OT receptors in mice. However, the regulatory elements that confer the specific expression patterns for the OT receptor gene in vivo are yet unknown.

One of the fundamental questions concerns the possible existence of OT receptor subtypes. Such subtypes have been suggested to be present, e.g., in the rat uterus, kidney, or brain, to explain differential pharmacological profiles or immunoreactivity patterns. Application of polymerase chain reaction methods and Southern analysis in several tissues known to possess OT binding activity failed to identify a gene encoding a further OT receptor subtype. However, the applied techniques only screen for genes with high homology to the uterine-type OT receptor, and therefore, a putative further OT receptor with low homology to the uterine-type OT receptor would have been kept undetected.

Receptor Structure B

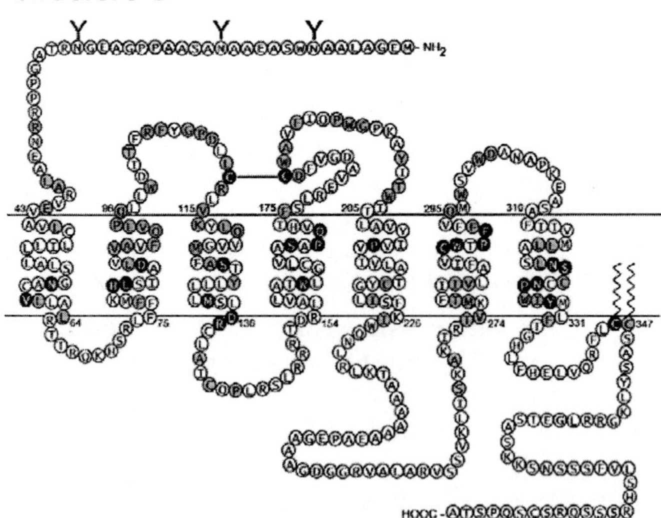

Figure: *Schematic structure of the human OT receptor with amino acid residues shown in one-letter code. The residues are marked in the same manner as in Fig. 3, i.e., residues conservative within the OT/vasopressin receptor subfamily are outlined in gray, and residues conservative for the whole G protein-coupled receptor superfamily are outlined in black. The putative* N-*glycosylation ("Y") and palmitoylation (at C346/C347) sites are marked.*

The OT receptor is a typical member of the rhodopsin-type (class I) GPCR family. The seven transmembrane α-helices are most highly conserved among the GPCR family members.

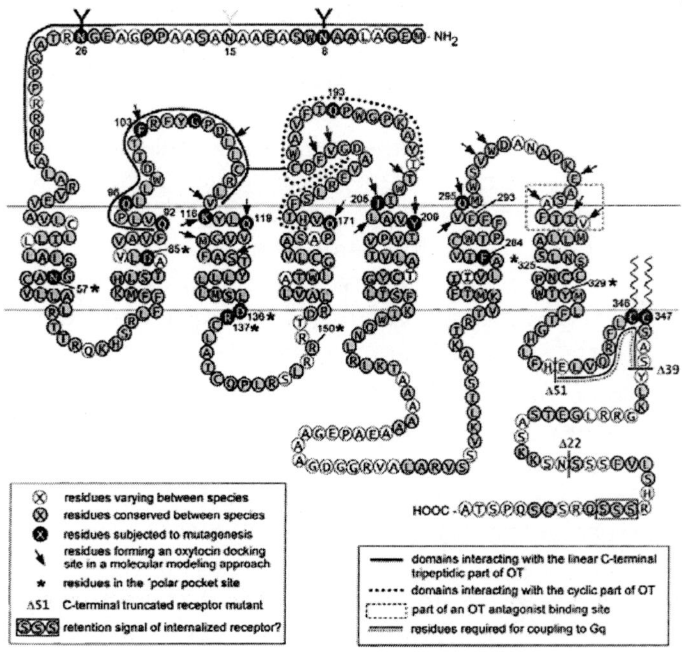

***Figure:** Schematic model of the human OT receptor indicating amino acid residues that are putatively involved in ligand-binding and associated signal transduction events (described in sect.III,* B–E*). The amino acid residues in gray solid circles are conserved (identical) between the OT receptors from different mammalian species (human, rhesus monkey, pig, bovine, sheep, rat, and mouse). Residues in open circles show interspecies variation. At these positions, an amino acid substitution may be tolerated through mammalian evolution without influencing the functional properties of the receptor. Residues in black solid circles have been subjected to mutagenesis. The glutamine and lysine residues highly conserved within the vasopressin/OT receptor family may partly define an agonist-binding pocket that is common to all the different subtypes of this receptor family. According to a molecular modelling approach, an OT docking site has been proposed (corresponding residues are marked by arrows). In the inactive receptor conformation, the highly conserved arginine may be constrained in a pocket that is formed by polar residues (indicated by asterisks). After agonist binding, this arginine side chain may be shifted out of the "polar pocket," thereby unmasking a G protein binding site. Receptor domains putatively interacting with OT, a peptide OT antagonist and* G_q*á are marked by lines.*

Conserved residues among the GPCRs may be involved in a common mechanism for activation and signal transduction to the G protein. On the basis of studies with model GPCRs, it is assumed that the switching from the inactive to the active conformation is associated with a change in the relative orientation of transmembrane domains 3 and 6, which then unmasks G protein binding sites. In the class I GPCR family, an Asp in transmembrane domain 2 and a tripeptide (E/D RY) at the interface of transmembrane 2 and the first intracellular loop are believed to be important for receptor activation. With respect to Asp-85, this was confirmed for the human OT receptor. When Asp-85 is exchanged by the residues Asn, Gln, or Ala, agonist binding and signal transduction of the receptor becomes impaired. Mutations at the conserved tripeptide motif DRY (DRC in case of the OT receptor) result in an either inactive or a constitutively active OT receptor.

The cysteine residues in the first and second extracellular loops are highly conserved within the GPCR family and are probably connected by a disulfide bridge. Two other well-conserved Cys residues reside within the COOH-terminal domain. Most likely, they are palmitoylated as demonstrated for the V2 receptor and other GPCRs and anchor the cytoplasmic tail in the lipid bilayer. However, for the V2 receptor as well as for the rat OT receptor, elimination of palmitoylation sites by mutagenesis failed to produce significant alterations in receptor function.

The OT receptor has two (mouse, rat) or three (human, pig, sheep, rhesus monkey, bovine) potential *N*-glycosylation sites (N-X-S/T consensus motif) in its extracellular NH_2-terminal domain. For the "core" OT receptor, a molecular mass of 40–45 kDa can be calculated on the basis of the amino acid sequence derived from the known cDNA sequences of several species. In photoaffinity labelling experiments using myometrial membranes obtained from guinea pig during late pregnancy, a 68- to 80-kDa protein was specifically labelled by a photoreactive OT antagonist developed by Elands et al.. Deglycosylation of the photolabelled receptor with endoglycosidase F gave rise to protein with 38–40 kDa. Similarly, in the same tissue, a 78-kDa protein was labelled by a photoreactive vasopressin analogue. In contrast, in membranes from rat mammary gland and rabbit amnion cells, photoreactive OT analogues specifically incorporated into a 65-kDa binding protein. It is possible that the different molecular masses for the myometrial versus the mammary gland and amnion OT receptor are due to differential glycosylation patterns. With the assumption of a mass of 10 kDa for a typical glycosylation core, all of the potential glycosylation sites could be occupied by glycosylation moieties. Recombinant deglycosylation mutants of the human OT receptor have been created by site-directed mutagenesis by exchanging Asp for Asn in positions 8, 15, and 26. The deglycosylated receptors were highly expressed in HeLa cells and showed unaltered receptor binding characteristics. Thus the receptor glycosylation appears not to be necessary for proper expression and has no effect on the functional properties of the receptor. Similar findings have been reported for other GPCRs, e.g., the vasopressin V2 receptor.

Ligand Binding Characteristics C.

The high homology of the nonapeptides of the evolutionary line isotocin-mesotocin-OT is also reflected in the high homology of the corresponding receptors. Accordingly, the mammalian OT receptors share the highest degree of sequence similarity with the toad mesotocin receptor (70%) (6) and the isotocin receptor of teleost fish (66%), whereas the sequence homologies with the vasopressin V1 (nearly 50%) and V2 receptors (40%) are significantly lower. About 100 amino acids (25%) are invariant among the 370-420 amino acids in the human receptors for vasopressin V2, V1a, V1b, and OT. The highest homology between the vasopressin/OT receptor types is found in the extracellular loops and the transmembrane helices. The NH_2 terminus and the COOH terminus have lower similarities, and the intracellular loops are the least of all conserved. Structural common features of the OT/vasopressin receptor family could play an important role in ligand/receptor recognition, e.g., the sequences FQVLPQ at the end of transmembrane domain 2, the sequences GPD (APD in mesotocin receptor) in the first extracellular loop, and DCWA (DCRA and DCWG in mesotocin and isotocin receptor) and PWG in the second extracellular loop.

For small molecules like catecholamines, the ligands bind in a cavity between the α-helical segments formed by transmembrane domains 3–6. Peptide ligands, on the other side, bind more superficially and also interact with extracellular loops and/or the NH_2-terminal domain. For the binding of the peptides OT and arginine vasopressin (AVP), residues located in the transmembrane domains as well as residues within extracellular domains are involved in ligand binding. Because the OT/vasopressin peptides as well as their receptors are well conserved, the ligand binding interaction should consist of both common and selective contact sites. As derived from molecular modelling in combination with mutagenesis studies, the agonist binding site for the vasopressin/OT peptides was proposed to be located in a narrow cleft delimited by the ringlike arrangement of the transmembrane domains. An equivalent position was described for the binding of cationic neurotransmitters. In the rat V1a receptor, the conserved Gln residues in the transmembrane domains 2, 3, 4, and 6 and a Lys residue localised in transmembrane domain 3 were replaced by Ala residues. All the receptor mutants had a decreased affinity for the agonists vasopressin, OT, and vasotocin. Because the corresponding Gln and Lys residues are highly conserved, it was proposed that the agonist-binding pocket is common to all the different subtypes of this receptor family.

Photoaffinity labelling of the first extracellular loop of the bovine vasopressin V2 receptor using a photoreactive lysine vasopressin analogue provided direct evidence for the involvement of the first extracellular loop in agonist binding. In the first extracellular loop, the homologous residues F103, Y115, and D115 in the human OT, V1a/V1b, and V2 receptor were found to be crucial for the determination of the ligand selectivity. For example, the mutation in the equivalent position (Y115F) of the rat V1a receptor led to a 19-fold increase of OT binding compared with the native receptor.

Molecular modelling of ligand binding interaction to the V1a receptor supported the view that the side chain of Arg-8 in AVP projects outside the transmembrane core of the receptor and could interact with Tyr-115 located in the first extracellular loop. Arg-8 in AVP is known to be necessary for its high-affinity binding to the V1a receptor. When Tyr-115 in the V1a receptor is replaced by an Asp and a Phe, the amino acids naturally occurring in the V2 and in the OT receptor subtypes, the agonist selectivity of the V1a receptor switches accordingly. The corresponding residue also determined the agonist specificity of the bovine and pig V2 receptor. Thus this residue certainly contributes to agonist selectivity. Additionally, by a peptide mimetic approach it was found that a synthetic dodecapeptide, which is homologous to the first extracellular loop of the human OT receptor, inhibits the binding of tritiated AVP to the human OT receptor. The second extracellular loop is also thought to be important for hormone binding of the human OT receptor, since it is conserved only within the nonapeptide receptor family.

The OT receptor has a weak ligand selectivity profile: hormones with the same cyclic part and either Arg-8 (in arginine vasotocin) or Leu-8 (in OT) are bound with the same affinity, whereby Ile-3 (in OT) in the cyclic hormone part contributes more to affinity than Phe-3 (in oxypressin). This indicates that the cyclic part of OT is more important in conferring binding selectivity for the OT receptor compared with the linear tripeptidic part of the hormone. Using chimeric "gain in function" V2/OT receptor constructs, Postina et al. demonstrated that the NH_2 terminus and the first and second extracellular loops were necessary for agonist binding and selectivity. In particular, the exchange of the NH_2 terminus of the V2 receptor for the corresponding first extracellular domain of the OT receptor resulted in a sixfold increase in binding affinity for OT.

Presumably, the NH_2 terminus of the OT receptor takes part in hormone binding and probably interacts with the hydrophobic leucyl residue in position 8 of the ligands. The NH_2-terminal domain and the first extracellular loop of the OT receptor are proposed to interact with the linear COOH-terminal tripeptidic part of OT, whereas the second extracellular loop of the OT receptor could be identified to interact with the cyclic hormone part. Concerning the binding for OT and AVP, the OT receptor is relatively unselective with only about 10-fold higher affinity of the receptor for OT. AVP acts as a partial agonist on the OT receptor. To elicit the same response as induced by OT, 100-fold higher concentrations of AVP are necessary. However, AVP becomes a full agonist when two aromatic residues of the OT receptor (Y209 and F284) are replaced by the residues F and Y present at equivalent positions in the vasopressin receptor subtypes. These two residues are therefore crucial for the response of the OT receptor to the partial agonist AVP.

Chimeric constructs encoding parts of the white sucker fish [Arg^8]vasotocin receptor and parts of the isotocin receptor have shown that the NH_2 terminus and a region spanning the second extracellular loop and its flanking transmembrane segments contribute to the affinity of the [Arg^8]vasotocin receptor. For the isotocin receptor from teleost fish, it has been shown that the sixth transmembrane helix and/or the fourth extracellular domain are involved in ligand binding.

Several studies indicate that the binding site of OT antagonists is different from the agonist binding site. Studies with chimeric receptors provided evidence that the binding site for the peptide OT antagonist was formed by the transmembrane helices 1, 2, and 7, with a major contribution to binding affinity by the upper part of helix 7. These regions did not participate in OT binding. Most mutations affecting agonist binding affinities have little effect on antagonist binding affinities.

Signal Transduction and G Protein Coupling D

GPCRs may also be constitutively active, in the absence of any agonist. This was first shown for the β-adrenergic receptor where mutations in the third intracellular loop, or simply overexpression of the receptor, resulted in constitutive receptor activation. The human V2 receptor mutant D136A and the human OT receptor mutant R137A represent such constitutively active receptors. Both positions are located within the conserved DRY motif (DRC in the OT receptor) at the cytoplasmic side of transmembrane domain 3. The invariably conserved Arg has been hypothesized to be constrained in a hydrophilic pocket formed by conserved polar residues in transmembrane domains 1, 2, and 7. Receptor activation was suggested to involve protonation of the Asp in this motif causing Arg to shift out of the polar pocket leading to cytoplasmic exposure of buried sequences in the second and third intracellular loops. In accordance with this hypothesis, mutating Asp in this motif resulted in increased agonist-independent activity of some receptors including the V2 receptor. Although for the V2 receptor as for some other receptors, mutations of the Arg residue within this motif result in uncoupled receptor forms, the human OT receptor mutant R137A possesses an increased basal activity. Thus, in this mutant, conformational constraints are released that normally stabilise the wild-type OT receptor in its inactive ground state. Activation of the OT receptor might occur similarly as proposed for the α_{1B}-adrenergic receptor, i.e., by the opening of a solvent-exposed site in the cytosolic domains that has been hypothesized to be involved in G protein recognition.

OT receptors are functionally coupled to $G_{q/11}\alpha$ class GTP binding proteins that stimulate together with $G\beta^3$ the activity of phospholipase C-β isoforms. This leads to the generation of inositol trisphosphate and 1,2-diacylglycerol. Inositol trisphosphate triggers Ca^{2+}release from intracellular stores, whereas diacylglycerol stimulates protein kinase C, which phosphorylates unidentified target proteins. Finally, in response to an increase of intracellular [Ca^{2+}], a variety of cellular events are initiated. For example, the forming Ca^{2+}-calmodulin complexes trigger activation of neuronal and endothelial isoforms of nitric oxide (NO) synthase. NO in turn stimulates the soluble guanylate cyclase to produce cGMP. In smooth muscle cells, the Ca^{2+}-calmodulin system triggers the activation of myosin light-chain kinase activity which initiates smooth muscle contraction, e.g., in myometrial or mammary myoepithelial cells. In neurosecretory cells, rising Ca^{2+} levels control cellular excitability, modulate their firing patterns, and lead to transmitter release. Further Ca^{2+}-promoted processes include gene transcription and protein synthesis.

In most cell systems studied so far, OT-induced intracellular Ca^{2+} increase is greater in the presence of extracellular Ca^{2+} than that in its absence. This suggests that OT has also effects on calcium influx through voltage-gated or receptor-coupled channels. The effect was nifedipine insensitive. OT was also shown to inhibit Ca^{2+}/Mg^{2+}-ATPase activity in sarcolemmal membranes from the rat uterine myometrium. This could sustain transient increases in intracellular Ca^{2+}concentrations and thereby prolong the effects of OT. In rat, guinea pig, and human myometrial cells, the OT-stimulated phosphoinositide hydrolysis was suggested to be mediated by pertussis toxin-sensitive and/or pertussis toxin-resistant G proteins. OT-stimulated GTPase and phospholipase C activities were attenuated by incubation with an antibody directed against the COOH termini of $G_q\alpha$ and $G_{11}\alpha$ in rat and human myometrial cells. In human myometrium, the coupling of OT receptors to a 80-kDa G protein with transglutaminase activity, termed G_h, was proposed from experiments with solubilised OT receptor-G protein ternary complexes. Phospholipase C-δ1 was suggested as the effector for this kind of signal transduction. In Chinese hamster ovary (CHO) cells expressing the rat OT receptor, OT stimulated increases in intracellular [Ca^{2+}], extracellular signal-related kinase-2 (ERK-2) phosphorylation, and PGE_2 synthesis. OT also induces PGE_2 synthesis in uterine endometrial and amnion cells. In cultured uterine myometrial cells, OT caused tyrosine phosphorylation of mitogen-activated protein (MAP) kinase through an islet-activating protein-sensitive G protein. Solubilisation experiments in combination with pertussis toxin sensitivity assays indicated that rat OT receptors can couple to both $G_{q/11}$ and $G_i$proteins in transfected CHO cells as well as in pregnant rat myometrium.

Which are the receptor domains conferring G protein specificity? Functional analysis of V1a/V2 hybrid receptors demonstrated that the second intracellular loop of the V1a receptor and the third intracellular loop of the V2 receptor each are required and sufficient for efficient coupling to $G_{q/11}$ and G_s, respectively. Cytoplasmic loops 2 and 3 are also proposed to be implicated in receptor G protein coupling for many other GPCRs. In case of the OT receptor, this appears to be more complex. Several intracellular domains of the receptor could be involved in the specificity and/or efficacy of coupling to $G_{q/11}$. This was concluded from the finding that various coexpressed intracellular receptor domains interfered with the OT-stimulated inositol phosphate production. Hoare et al. provided evidence that proximal parts of the COOH terminus of the rat OT receptor are required for coupling to G_q. Whereas OT receptors with COOH-terminal truncations of 22 and 39 residues showed no effect on receptor function, the OT receptor lacking 51 COOH-terminal residues revealed an interesting phenotype: OT-induced intracellular [Ca^{2+}] transients could be produced, although the phosphoinositide pathway was apparently not activated. However, it remained unclear which signals could mediate this Ca^{2+}release from intracellular stores. The Δ51 mutant receptor had a reduced affinity for OT and was uncoupled from G_q- mediated pathways. A coupling of this receptor to $G_i$was concluded, since the OT-induced Ca^{2+} transients were sensitive to pertussis toxin and to a $G\beta^3$ sequestrant. Because the Δ39 mutant was still able to couple to both $G_{q/11}$ and G_i, the sequence comprising the residues 339–350 of the rat OT receptor is required for interaction with $G_{q/11}$, but not G_i. Because of the high conservation of the COOH terminus of the OT receptor between various species, similar signal transduction mechanisms may also occur for OT receptors from species other than rats. It is possible that the fidelity of receptor G protein interaction is decreased when OT receptors are strongly upregulated, e.g., in myometrium near term. Moreover, it is known that phosphorylation of GPCRs not only induce their desensitisation but may also modify their coupling specificity.

Receptor Internalisation and Downregulation E

When receptors are persistently stimulated with agonists, they desensitise. This process can occur by numerous mechanisms operating at the transcriptional, translational, and protein levels. Rapid, i.e., within seconds to minutes, homologous desensitisation of GPCRs consists of two steps, phosphorylation and subsequent arrestin binding. The receptor uncouples from G proteins and undergoes endocytosis, internalisation, or sequestration. Receptor sequestration is viewed as an early step in the downregulation of receptors that occurs after prolonged (hours to days) agonist stimulation and that may either end in degradation within lysosomes or in recycling back to the plasma membrane. These processes have been best studied for the adrenergic receptors. Birnbaumer and co-workers have analysed the internalisation process for the vasopressin V1a and V2 receptors in some detail.

Like most other GPCRs, OT receptors may undergo rapid homologous desensitisation following persistent agonist stimulation.

Within 5–10 min. after agonist stimulation, >60% of the human OT receptors expressed in HEK 293 fibroblasts were internalised (Gimpl and Fahrenholz, unpublished data), similar to as found for the human V2 receptor expressed in the same system and in LLC-PK_1 cells.

Internalisation of OT receptors occurs mainly by a clathrin-dependent pathway. But when stably expressed in HEK 293 cells, a fraction of OT receptors (10–15% of total) is localised in caveolae-like membrane microdomains.

The internalisation mechanism for this receptor population has not been examined. The internalised OT receptor is not recycled back to the cell surface (Gimpl and Fahrenholz, unpublished data). This indicates that the OT receptor behaves more like the V2 receptor and unlike the V1a receptor that rapidly recycles back to the cell surface.

Using mutagenesis experiments and chimeric receptor constructs, Innamorati et al. identified a serine cluster (Ser-362 to Ser-364) in the COOH-terminal tail of the V2 receptor acting as a retention signal for the internalised V2 receptor. The human OT receptor contains 17 potential phosphorylation sites including two serine clusters in its COOH terminus. One may speculate that these clusters also contribute to prevent the recycling of the internalised OT receptor.

Exposure of human myometrial cells to OT for up to 20 h resulted in an almost 10-fold reduction in OT binding capacity. Although the total amount of OT receptor protein appeared not to be affected by OT treatment for up to 48 h, the OT receptor mRNA was reduced, which may be due to transcriptional suppression and/or destabilisation of mRNA. When HEK 293 cells expressing the human OT receptor were treated for 18 h with high (¼M) concentrations of OT, 50% of the initial binding capacity remained at the cell surface. In WRK1 cells, OT was able to induce a desensitisation of the vasopressin (VP) receptors when present for 18 h.

Up- or downregulation of receptors could also be affected by yet ill-defined cross-talk mechanisms. Evidence for a cross-talk between the corticotropin-releasing hormone (CRH) and OT signal transduction pathways was provided in human myometrial cells at term. Furthermore, stimulation of β_2-adrenergic receptors causes heterologous upregulation of OT receptors in the nonpregnant estrogen-primed rat myometrium. In this system, a threefold increase in OT receptor mRNA, an 100% rise in receptor binding, and an augmented contractile response of isolated uterine strips to OT were observed.

Effects of Steroids F

Cholesterol 1: Both solubilised and membrane-associated OT receptors require at least two essential components for high-affinity OT binding: divalent cations such as Mn^{2+}or Mg^{2+} and cholesterol. Compared with many other GPCRs, the GTP sensitivity of the agonist binding to the OT receptor is rather modest. All attempts to purify functional OT receptors have been unsuccessful to date. With the use of 3-[(3-cholamidopropyl)dimethylammonio]-2-hydroxy-1-propanesulfonate (CHAPSO) as detergent, it is possible to solubilise functional OT receptors from different sources. However, a common observation is that following solubilisation, OT receptors lose characteristic binding properties, the affinity for OT becomes lower, and/or additional low-affinity state receptors appear in the extract. Unfortunately, low-affinity [e.g., dissociation constant (K_d) >10–50 nM] receptor populations cannot be characterised adequately by conventional radioligand binding assays. The necessity to separate free from bound ligand concomitantly leads to a dissociation of low-affinity ligand-receptor interactions. Solubilisation with CHAPSO is known to lead to a substantial cholesterol depletion of the soluble extract, and we could demonstrate that substitution with cholesterol markedly enhanced the OT binding of soluble OT receptors. This became first evident in reconstitution of soluble OT receptors using liposomes of defined composition. A saturable high-affinity OT binding was obtained only with liposomes that contained a critical amount of cholesterol.

Moreover, when OT receptors were expressed in insect cells, which naturally have plasma membranes with low cholesterol content, the receptors are mainly in a low-affinity state (K_d >100 nM). After addition of cholesterol to the culture medium, a fraction of OT receptors is converted from a low- to a high-affinity state (K_d 1 nM). The low-affinity state was identified as a physiological active receptor state, and the conversion of the affinity states to each other is, at least to a certain degree, reversible. The interaction of cholesterol with OT receptors is of high specificity and is not due to mere changes of membrane fluidity. Furthermore, cholesterol stabilises both membrane-associated and solubilised OT receptors against thermal denaturation. Taken together, our data suggest a direct and cooperative molecular interaction of cholesterol with OT receptors. Cholesterol acts as an allosteric modulator and stabilises the receptor in a high-affinity state for agonists and antagonists. In many but not in all cell systems, populations of high- and low-affinity OT receptors have been observed. This could reflect uneven cholesterol distributions within the plasma membrane of these cells.

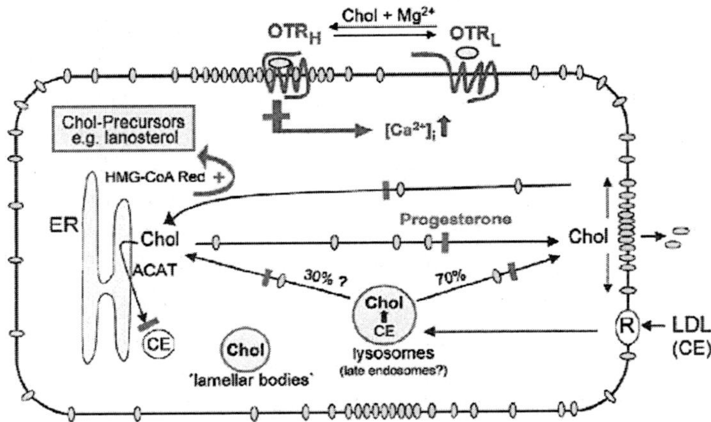

Fig. 6. Schematic model of nongenomic inhibitory effects of progesterone. Progesterone inhibits both the signal transduction of Gq couplled receptors (as shown here for the OT receptor) and the intracellular trafficking of cholesterol. Principally, eukaryotic cells can obtain the required cholesterol (Chol, gray ellipses) by two sources: endogenously by de novo synthesis of cholesterol and exogenously by uptake of cholesteryl ester (CE)-rich low-density lipoprotein (LDL) particles via receptor-mediated (R) endocytosis. De novo synthesized cholesterol first arrives at cholesterol-rich domains in the plasma membrane (caveolae and/or "lipid rafts") that may function as cholesterol "sorting centres" within the plasma membrane, where most of the cellular cholesterol resides. Progesterone blocks several intracellular transport pathways of cholesterol (red bars) except for the LDL receptor-mediated uptake of cholesterol. Moreover, cholesterol esterification does not occur in the presence of progesterone, presumably due to the lack of cholesterol substrate for acyl-CoA:cholesterol acetyltransferase (ACAT). As a consequence, unesterified cholesterol accumulates in lysosomes (or late endosomes) and lysosome-like compartments (designated as "lamellar bodies") (marked by red background). The key enzyme for the cholesterol de novo synthesis, 3-hydroxy-3-methylglutaryl-CoA reductase (HMG-CoA Red), is stimulated in the presence of progesterone (red arrow), but the cholesterol biosynthesis stops at the level of precursors (e.g., lanosterol). Enzymes involved in the conversion of cholesterol precursors reside in the endoplasmic reticulum (ER), and progesterone most likely prevents sterol precursors localised in the plasma membrane from reaching the ER-resident enzymes, thereby preventing their conversion to cholesterol. Overall, progesterone induces a state of cholesterol auxotrophy. However, after progesterone withdrawal, the accumulated precursors will be rapidly converted to cholesterol. Thus cells will become overloaded with cholesterol for a certain period of time, after which the cholesterol homeostasis will be reestablished. We hypothesize that these reversible progesterone-induced changes of the cholesterol trafficking could have a strong influence on signal transduction processes, particularly in case of the OT receptor (OTR_H and OTR_L, high-affinity and low-affinity OT receptor, respectively; receptor in blue; OT in yellow).

We would expect that high-affinity state OT receptors are preferentially localised in cholesterol-rich subdomains of the plasma membrane. Likewise, receptor heterogeneity with respect to affinity states should be highest in cell systems with abundant cholesterol-rich domains such as rafts or caveolae structures, e.g., in myometrial cells at term. In fact, we recently provided evidence for a partial enrichment of high-affinity OT receptors in cholesterol-rich plasma membrane domains in HEK 293 fibroblasts stably expressing the human OT receptor.

Divalent metal ions like Mg^{2+} are long known to increase the response of target cells to OT and to shift the dose-response curve to the left. Thus addition of Mg^{2+} was found to increase both the OT binding capacity and the affinity state of the OT receptor. This is surprisingly similar to what cholesterol does. In addition, Mg^{2+} has been proposed to display its effect on the OT receptor interaction by influencing positive cooperativity. Mg^{2+} increases the potency of OT analogues in stimulating uterine contractions whereby the effects of Mg^{2+} were observed to be inversely related to the potency of the peptide.

Relatively inactive peptides like 7-glycine OT became significantly more potent when the Mg^{2+} concentration bathing the uterine smooth muscle in vivo was increased from 0 to 0.5 mM. Conclusively, cholesterol and Mg^{2+} are essential allosteric modulators of the OT receptor and may be involved in the regulation of OT-mediated signalling functions.

Do these allosteric modulators play a role for the regulation of OT-related physiological processes? Some reports suggest that particularly in reproductive tissues, the cholesterol concentrations may be highly dynamic. With the use of freeze-fracture cytochemistry with the cholesterol-binding filipin, marked increases in cholesterol have been found in rat uterine epithelial cells at the time of blastocyst implantation. In the human placental syncytiotrophoblast basal membrane, Sen et al. observed a steady decrease in cholesterol-to-phospholipid ratio in correlation with an increase in membrane fluidity during placental development. At term however, the cholesterol-to-phospholipids ratio in syncytiotrophoblast membranes was found to be increased compared with the cholesterol-to-phospholipid ratio in early placentas. Moreover, cholesterol-enriched caveolae structures are a conspicuous feature in the rat myometrium at term. We have provided evidence that cholesterol can modulate receptor function by both changes of the membrane fluidity and direct binding effects, e.g., in case of the OT receptor. Plasma membranes with lowered cholesterol content showed a decreased capacity (B_{max}) of binding sites and/or a decreased affinity (K_d) of ligand-receptor binding. Interestingly, Lopez et al. reported that pregnancy in humans was associated with increases in both density and affinity of OT receptors. To draw further conclusions, correlation studies are required using tissues in which both the membrane cholesterol content and the OT receptor activity will be measured at the same time.

Progesterone 2

Progesterone is considered to be essential to maintain the uterine quiescence. Grazzini et al. recently postulated that progesterone specifically binds to the rat OT receptor with high affinity (K_d 20 nM) and thereby inhibits the receptor function. In case of the human OT receptor, a direct inhibitory interaction [inhibitory constant (K_i) 30 nM] with a progesterone metabolite, 5β-pregnane-3,20-dione, has been reported by the same authors. They claimed that progesterone could act as a negative modulator of the OT receptor and thus offered a plausible mechanism of how progesterone could contribute to uterine quiescence. However, these findings could not be reproduced in several other laboratories including our own. Instead, we found that high concentrations of progesterone (>10 ¼M) attenuated or blocked the signalling of several GPCRs, including the OT receptor. The progesterone effects occurred within minutes, were reversible, and could not be blocked by a protein synthesis inhibitor. Overall, the action of progesterone was more cell type specific than receptor specific. The progesterone doses that are required to affect the signalling function of receptors are much higher than the progesterone levels found in plasma or in nonsteroidogenic tissues such as the myometrium. In steroidogenic tissues, however, huge amounts of progesterone have been measured.

In human corpus luteum, progesterone concentrations reached peak levels of 25 ìg/g tissue shortly after ovulation and in the early luteal phase. These values are within the range of the progesterone concentrations that were effective in our study. Thus, in steroidogenic cells as well as in their environment, progesterone might nongenomically influence the signalling of receptors. The molecular mechanisms underlying this progesterone action are not understood. A well-known progesterone binding protein is the multidrug resistance P-glycoprotein.

In addition to their role in detoxification, P-glycoproteins are involved in intracellular cholesterol transport. It is known that progesterone markedly interferes with the intracellular transport (and metabolism?) of cholesterol. At concentrations in the micromolar range, it inhibits both the cholesterol esterification and the transport of cholesterol to and from the plasma membrane. In particular, progesterone reduces the cholesterol pool residing in caveolae . Paradoxically, at the same time, progesterone stimulates the activity of 3-hydroxy-3-methylglutaryl (HMG)-CoA reductase, the key enzyme of de novo cholesterol biosynthesis. Hence, cholesterol precursors like lanosterol begin to enrich in the membranes of the cell. As mentioned above, the OT receptor needs a cholesterol-rich microenvironment to become stabilised in its high-affinity state. Because the cholesterol precursors, particularly lanosterol, are completely inactive to support the OT receptor in its high-affinity state, the responsiveness of the OT system may not be fully operative during the continuous presence of high progesterone concentrations.

According to this scenario, progesterone withdrawal would restore the cholesterol transport so that the highly enriched amounts of cholesterol precursors would now become rapidly converted to cholesterol. This would lead to a sudden rise of cholesterol and should push the responsiveness to OT since low-affinity OT receptors could now be converted into their high-affinity state. According to this postulated mechanism, progesterone could affect the signalling of all those receptors that are functionally dependent on cholesterol. It is important to note that the nongenomic actions of progesterone including its influence on the cholesterol transport require progesterone concentrations in the micromolar range. This suggests that the described effects may be limited to the steroidogenic tissues and to their environment. Most likely, progesterone acts in these tissues via both genomic and nongenomic pathways together with other steroids to control receptor activity.

The Peripheral Oxytocin System

Female Reproductive System A

Uterus 1: The pregnant uterus is one of the traditional targets of OT. OT is one of the most potent uterotonic agents and is clinically used to induce labour. Accordingly, the development of highly specific OT antagonists may be of therapeutic value for the prevention of preterm labour and the regulation of dysmenorrhea.

The OT gene was found to be expressed in the rat uterine epithelium at term. The estrogen-induced elevation of OT mRNA levels was restricted to 3 days and reached peak levels that exceeded hypothalamic OT mRNA levels by a factor of 70. In rats, OT gene expression was shown to be present in placenta and amnion and in humans in amnion, chorion, and decidua. However, in most studies, significant increases of OT before the onset of labour have not been detected, neither in maternal plasma nor in intrauterine tissues. On the other hand, some findings suggest that there is a relationship between the pattern of OT secretion and advancing pregnancy. In rhesus monkeys, it was shown that maternal but not fetal OT concentrations were positively correlated with nocturnal uterine activity and progressively increased during late pregnancy and delivery.

Around the onset of labour, uterine sensitivity to OT markedly increases. This is associated with both an upregulation of OT receptor mRNA levels and a strong increase in the density of myometrial OT receptors, reaching a peak during early labour. This has been demonstrated both in the rat and in the human species, in which receptor levels rise during early labour to 200 times that in the nonpregnant state. Thus, at the onset of labour, OT can stimulate uterine contractions at levels that are ineffective in the nonpregnant state. After parturition, the concentrations of OT receptors rapidly decline. In rats, the uterine OT receptor mRNA levels decreased more than sevenfold within 24 h. Possibly, the downregulation of the OT receptors may be necessary to avoid unwanted contractile responses during lactation when OT levels are raised.

Gonadal steroids play an important role in the regulation of uterine OT receptors. In the days preceding birth, the ratio of plasma progesterone to estrogen falls. These changes in the steroid concentrations may occur in most mammals. At least in humans, progesterone withdrawal has not been determined. The steep drop of circulating progesterone occurs after placental delivery. Alteration of sex steroid metabolism in the fetoplacental unit appears to occur in women and primates. As shown in bovine and in sheep, maturation of the fetal hypothalamus leads to an increased secretion of CRH, which in turn stimulates the pituitary to secrete ACTH. Subsequently, ACTH stimulates the fetal adrenal to release cortisol.

Additionally, OT-induced contractures of the myometrium in late pregnancy could lead to temporary decreases in blood flow and transient episodes of fetal hypoxia, which also may provoke a fetal stress response. Cortisol then increases the activity of the key enzyme cyctochrome *P*-450c17α, which promotes placental pregnenolone turnover into estrogens. In primates, CRH is additionally synthesized by the placenta and stimulates the fetal adrenal to secrete dehydroepiandrosterone sulfate (DHEAS), the precursor of placental estrogen.

In each case, the ratio of estrogen to progesterone increases in the maternal plasma concomitantly with increases in the synthesis of connexin-43, formation of gap junctions, increased production of prostaglandins from intrauterine tissues, and an upregulation of OT receptors. Finally, the uterine quiescence that was maintained by the high progesterone level ceases, and parturition can occur. Upregulation of OT receptors and an increased expression of gap junctions also occur in the hours or days preceding human labour onset. Obviously, estrogens and progesterone act in opposing fashion on the function, expression, and/or regulation of OT receptors. The mechanisms for the sudden and sometimes unpredictable responsiveness to OT of the myometrium is still mysterious. As Kimura pointed out, the reaction of the uterus to OT in the same patient could vary from day to day. To induce labour at term, in some patients, OT is completely ineffective even at high doses, whereas in others a minimal dose can induce hypertonus of the uterus.

However, despite the striking steroid dependence of the OT system, the promoter of the human OT receptor gene does not contain a classical steroid responsive element. Several observations indicated that progesterone promotes uterine relaxation and inhibits the function of the OT receptor system by both genomic and nongenomic mechanisms. Even in the absence of protein synthesis, progesterone induces a reduction in uterine OT binding. Moreover, progesterone-mediated downregulation of OT receptors was not accompanied by a decrease in OT receptor gene expression. The mechanisms of how progesterone acts nongenomically are still unknown but are of fundamental importance.

The increase in OT receptors before labour is not confined to the myometrium. It is also found in the decidua. During the course of parturition, the OT receptor gene expression in human chorio-decidual tissue was fivefold increased. In the decidua, OT has a separate action of stimulating the release of prostaglandin $PGF_{2\alpha}$. At the end of pregnancy, an increased secretion of $PGF_{2\alpha}$ drives luteolysis and thus leads to progesterone withdrawal and labour initiation in rodents. Mice lacking the gene encoding the $PGF_{2\alpha}$ receptor developed normally but were unable to deliver normal fetuses at term. These knock-out mice neither responded to OT nor did they show an induction of OT receptors. Furthermore, the normal decline of serum progesterone concentrations that precedes parturition did not occur. Ovariectomy restored induction of OT receptors and permitted successful delivery in the $PGF_{2\alpha}$ receptor-deficient mice. This indicates that $PGF_{2\alpha}$ acts upstream of OT to induce luteolysis, i.e., production of progesterone. The OT system may be regarded as a key regulator of parturition, with the induction of OT receptors as a trigger event. However, in OT-deficient mice, parturition remained unaffected. This suggests that at least in mice, uterotonins other than OT are (additionally?) operative as myometrial contractants to initiate labour. A recent study with mice deficient in both OT and cyclooxygenase-1 shed more light for OT's role at the onset of labour. Cyclooxygenase-1 is involved in the synthesis of prostaglandins, and as expected, mice lacking this enzyme showed reduced levels of $PGF_{2\alpha}$, impaired luteolysis, a less pronounced fall in progesterone in late gestation, and finally a delayed initiation of labour.

Surprisingly, mice deficient in both OT and cyclooxygenase-1 initiated labour at the normal time. Thus OT and prostaglandins had apparently opposing actions on the onset of murine parturition: a luteotrophic function of OT versus a luteolytic action of prostaglandins. So, OT stabilises the progesterone synthesis before parturition. For the normal timing of labour it might be essential to generate sufficient $PGF_{2\alpha}$ to overcome the luteotrophic action of OT in late gestation. These findings might explain why the upregulation of OT receptor occurs without an increased expression of OT shortly before the onset of labour. The induction of uterine OT receptor expression allows OT to act as a potent uterotonic agent, whereas an increase in OT expression could prolong gestation due its trophic effects on the corpus luteum. Fuchs et al. reported that OT may also be involved in induction of cyclooxygenase-2 expression and $PGF_{2\alpha}$ release at term and during parturition in cows.

In ruminants, the endometrial OT receptor exhibits a cycle-dependent regulation and plays a crucial role in reproduction. In the bovine endometrium, OT receptor mRNA levels can exceed the levels in the myometrium at the same stage of the cycle. Within 2–3 days around estrus, the endometrial OT receptors increase to a level similar to that observed at term of pregnancy. For the remainder of the cycle, OT receptor concentrations are almost undetectable. Pulsatile secretion of endometrial $PGF_{2\alpha}$ is stimulated by OT during late diestrus in domestic ruminants and results in regression of the corpus luteum leading to the onset of a new estrous cycle. Thus the uterus influences the length of the luteal phase via the expression of the endometrial OT receptor. The administration of an OT antagonist or the continuous administration of OT, which downregulates the OT receptor, delays the onset of luteolysis and lengthens the cycle. A steep rise in endometrial OT receptors also precedes the onset of labour in ruminants. In cows, the receptor density increases almost 200-fold and remains at high levels during labour. Suppression of the endometrial OT receptor gene expression was shown to be caused by antiluteolytic proteins, e.g., interferon-[3] secreted by the trophoblast of the developing blastocyste. In primates, the endometrium is not essential for luteolysis, since hysterectomy does not abolish cyclic ovarian function. In humans, Northern blotting revealed the presence of OT receptor mRNA in the pregnant myometrium, in the endometrium, and in the ovary. In the nonpregnant human uterus, the OT receptor mRNA is mainly expressed in the glandular epithelial cells of the endometrium with highest expression level occurring at ovulation. However, its physiological role is unclear.

One of the highest concentrations of OT receptors (10 pmol/mg protein) has been measured in rabbit amnion at term. At the end of gestation, rabbit amnion OT receptors increase more than 200-fold. This receptor upregulation is associated with the release of PGE_2 from cultured amnion cells and suggests an important role for the amnion and OT in the initiation of labour in rabbits. In the same system, cortisol and cAMP caused a marked physiological upregulation of OT receptors as well as an increased PGE_2 response to OT. Addition of cortisol to amnion cells increased OT-stimulated PGE_2 release almost 100-fold, whereas the combination of forskolin and cortisol increased the PGE_2 response to OT 5,600 times. OT receptors in human amnion were also associated with PGE_2 release and were found to be upregulated in early and advanced labour, albeit to a much lower degree. In contrast, no OT binding sites were observed in amniotic/chorionic or placental membranes from rats.

Many studies have supported the view that OT is synthesized in a paracrine system operating within human intrauterine tissues. This includes the fetal membranes amnion and chorion and the maternal decidua. These tissue layers are continuously bathed in amniotic fluid and could transmit signals of maternal or fetal origin to the myometrium. Because fetal membranes are capable of steroidogenesis, they may contribute to local changes in the estrogen-to-progesterone ratio. Although there is evidence that during late gestation OT gene expression is upregulated in intrauterine tissues, the OT peptide itself was not found at concentrations significantly higher than in the circulating blood. This could be due to the high activity of oxytocinases, but that still remains to be shown. The possibility of local, paracrine effects of OT was also suggested to be present within the intrauterine tissues of the rat and bovine. In the pregnant cow, OT mRNA levels were found to be very low except for the corpus luteum. After the onset of labour, both OT gene and OT peptide were expressed at significant levels in the corpus luteum. This rescue of luteal OT at term could act to supplement the circulating hormone of pituitary origin. Thus OT may act primarily as a local mediator and not as a circulating hormone during parturition. In an autocrine/paracrine system within the uterus, significant changes of OT, prostaglandins, and sex steroids could occur without being reflected in the maternal circulation.

Ovary and Corpus Luteum 2

In several species, the ovary has been shown to contain OT and may be a site of local OT production. In the marmoset monkey *(Callithrix jacchus)*, in vivo and in vitro studies suggested that OT is a follicular luteinization factor. After human chorionic gonadotropin treatment, almost all granulosa cell layers in antral follicles showed immunoreactivity for both OT and OT receptor. OT was produced only by granulosa cells derived from preovulatory follicles, and after application of OT, only the granulosa cells cultured from preovulatory follicles elicited an increase in progesterone production. Functional OT receptors have been detected in bovine granulosa cells, suggesting that OT may be an autocrine factor during follicular growth. OT also increased the rate of mouse blastocyst development and might therefore play some role in the early stage of development of fertilized oocytes. Both OT and OT receptor genes are expressed in human cumulus cells surrounding the oocytes. Thus local OT may participate in fertilization and early embryonic development in humans.

The interaction of neurohypophysial OT with endometrial OT receptors evokes the secretion of luteolytic pulses of uterine $PGF_{2\alpha}$. McCracken et al. have postulated the concept of a central OT pulse generator that functions as pacemaker for luteolysis.

According to this concept, the uterus transduces hypothalamic signals in the form of episodic OT secretion, into luteolytic pulses of uterine $PGF_{2\alpha}$. In ruminants, luteal OT is released synchronously by uterine $PGF_{2\alpha}$ pulses. Thus a positive-feedback loop is established that amplifies neural OT signals.

The onset of this feedback loop and episodic $PGF_{2\alpha}$ secretion is controlled by the appearance of OT receptors in the endometrium. During pregnancy, the developing conceptuses must prevent endometrial $PGF_{2\alpha}$ from being released into the uterine vasculature to prevent corpus luteum regression and progesterone withdrawal.

The continued progesterone secretion is required for establishment and maintenance of pregnancy.

Luteal cells produce OT, but they can also be target cells for OT action. OT receptors have been characterised on both steroidogenic cell types of the corpus luteum, small and large cells. In several species, OT release was highest in the young corpus luteum. In the pig, the density of OT receptors was also highest during the early luteal phase, suggesting that autocrine/paracrine actions of OT may occur primarily in the young corpus luteum. The effects of OT on cultivated luteal cells were also strongly dependent on the age of the corpus luteum. In the pig, a function for OT in the ovary has been suggested in both a luteotrophic and luteolytic context. Luteal OT and progesterone release occurs in tightly coupled pulses. In vivo, OT and $PGF_{2\alpha}$ stimulate estradiol and progesterone release, and estradiol itself further stimulates progesterone release. Thus there may be an intraluteal circuit that involves paracrine effects of estradiol, OT, and $PGF_{2\alpha.}$

Male Reproductive Tract B

Testis 1

In several species, a pulse of systemic OT, presumably of hypothalamic origin, appears to be associated with ejaculation. The systemic hormone could act peripherally stimulating smooth muscle cells of the male reproductive tract, but could also reflect central effects in the brain modulating sexual behaviour.

OT has been identified in the testes from various mammalian species, and a mesotocin-like peptide has been demonstrated in the testes of birds and marsupials. There is evidence that OT is made locally within the testis, and possibly also the epididymis and prostate. Concerning the localisation of the OT system in the male reproductive tract, species-specific differences are important to note. For example, mice do not have detectable levels of OT mRNA in their testes, whereas cattle have relatively high levels of testicular OT mRNA. In contrast to the eutherian mammals, in the tammar wallaby (*Macropus eugenii*), the mesotocin receptor gene and protein, which are highly homologous to the eutherian OT receptor, are expressed in the prostate gland, but not in the testis. In the human, the complete OT system appears to be present in testis, epididymis, and prostate.

In the rat testis, OT concentrations are higher than those in the peripheral circulation. OT gene transcripts could be detected using

PCR analysis but not by Northern hybridisation. Within the rat testes, OT is present in the interstitial Leydig cells, the cells which provide the main source of testosterone in the male. Testicular OT levels are not constant but vary in dependence of the level of gonadotrophins and the activity of the seminiferous epithelium. OT production is increased in vitro by luteinizing hormone (LH), whereas testosterone itself has no effect on the secretion of OT. Notably, OT was only detectable during active spermatogenesis. Primarily two functions have been ascribed to testicular OT, namely, the regulation of seminiferous tubule contractility and the modulation of steroidogenesis. In the testis, the seminiferous tubules are surrounded by smooth muscle-like cells, the myoid cells. OT has been shown to enhance the contractility of the tubules; therefore, the responsiveness to OT was higher at a certain stage of the spermatogenic cycle, around the time when the sperm are shed into the lumen.

Obviously, OT promotes the spermiation and the subsequent transport of the immotile spermatozoa to the epididymis. However, there is no clear evidence for the expression of OT receptors on myoid cells, the cells which are mainly responsible for the contractile activity of the tubules. Because Sertoli cells exhibit the components of a local OT system under some circumstances, they may also be involved in the contractile activity of seminiferous tubules. On the other hand, it was suggested that the contractility effects are mediated by V1a receptors that are present in testes at severalfold higher densities compared with OT receptors.

Uterine-type OT receptors have been identified in the interstitial spaces in the rat testis consistent with binding to Leydig cells. Also in the male marmoset monkey (*Callithrix jacchus*), OT and its receptor have been detected predominantly within the Leydig cells. The Leydig cells drive spermatogenesis via the secretion of testosterone, which acts on the Sertoli and/or peritubular cells to create an environment that enables normal progression of germ cells through the spermatogenic cycle. Daily subcutaneous injections of OT led to an increase in the plasma and testicular levels of testosterone. In contrast, continuous administration of the peptide into the testis, whether by implants or in the transgenic mouse which overexpresses the bovine OT gene in its testes, produced a decrease in testosterone but an increase in dihydrotestosterone concentrations. Although in the testis the predominant androgen is testosterone, elsewhere in the male reproductive tract testosterone acts as a prohormone and is converted to its active metabolite dihydrotestosterone by the enzyme 5αreductase. OT was shown to increase the activity of 5α-reductase in both testis and epididymis and may thus have an autocrine/paracrine role modulating steroid metabolism in these tissues.

Prostate Gland 2

The prostate is an androgen-dependent gland. Testosterone enters the cell and is converted to dihydrotestosterone, which then modulates growth and prostatic functions. OT is present in the prostate at concentrations higher than in the plasma and can increase the resting tone of prostatic tissue from guinea pig, rat, dog, and human. OT also evoked contractile activity on mammalian prostates in vitro. Thus it was suggested that OT is involved in the contraction of the prostate and the resulting expulsion of prostatic secretions at ejaculation.

OT may also be involved in the pathophysiology of benign prostatic hyperplasia. In humans, benign prostatic hyperplasia occurs spontaneously, is often associated with bladder outlet obstruction, and affects 50% of men over the age of 60. However, obstructive symptoms may not only be caused by an enlargement of the gland but also by an increase in smooth muscle tone. So, in the nonoperative management of prostatic hyperplasia, α_1-adrenergic blockers are presently used. Because OT is even more potent at increasing smooth muscle tone than norepinephrine, OT antagonists may be potentially useful in the treatment of this disease. OT can stimulate growth of the prostate in the rat, particularly the mitotic activity in the glandular epithelium. OT treatment elevated the testosterone levels of plasma, testes, and prostate. Conversely, prostatic OT concentrations are found to be decreased by testosterone and increased after castration. In rats, OT treatment only transiently increased 5α-reductase activity in the prostate, consistent with the short-lived increase of dihydrotesterone in this tissue. These observations suggest that local feedback mechanisms act to control prostatic levels of dihydrotestosterone and prostatic growth. The findings in rats were somewhat different to those in dogs, which unlike rats can spontaneously develop prostatic hyperplasia. Interestingly, old dogs with established prostatic hyperplasia showed raised OT levels accompanied by elevated 5α-reductase activity within their prostatic tissues. It is hypothesized that OT acts as a paracrine factor to regulate cell growth via its influence on the enzyme 5α-reductase.

Mammary Tissues C

Milk Ejection 1: One of the classical roles assigned to OT is milk ejection from the mammary gland. The secretion of the mammary glands is triggered when the infant begins to suck on the nipple. The stimulation of tactile receptors at that site generates sensory impulses that are transmitted from the nipples to the spinal cord and then to the secretory oxytocinergic neurons in the hypothalamus. These neurons display a synchronised high-frequency bursting activity, consisting of a brief (3–4 s) high-frequency discharge of action potentials recurring every 5–15 min. Each burst leads to massive release of OT into the bloodstream by which OT is carried to the lactating breasts. There it causes contraction of the myoepithelial cells in the walls of the lactiferous ducts, sinuses, and breast tissue alveoli. In humans, within 30 s to 1 min. after a baby begins to suckle the breast, milk begins to flow. This process is called milk ejection or milk let-down reflex and continues to function until weaning.

As mentioned above, central oxytocinergic neurons are decisive components for the initiation and maintenance of successful lactation. So, the reflex that elicits OT release in women is activated before the tactile stimulus of suckling occurs and is related to such factors as the baby crying. The essential role of OT for the milk let-down reflex has been confirmed in OT-deficient mice. In fact, the major deficit of these female OT knock-out mice was the failure to nurse their offspring. In addition, the presence of OT in conjunction with continued milk removal was also required for postpartum alveolar proliferation and mammary gland function. Alveolar density and mammary epithelial cell differentiation at parturition were similar in wild-type and OT-deficient dams. However, within 12 h after parturition, 2% of the alveolar cells in wild-type dams incorporated DNA and proliferated, whereas no proliferation was detected in OT-deficient dams. Continuous suckling of pups led to the expansion of lobulo-alveolar units in wild-type but not in OT-deficient dams. Despite suckling and the presence of systemic lactogenic hormones, mammary tissue in OT-deficient dams partially involuted. Continuously elevated OT concentrations such as those during infusion or during normal milking are also necessary for complete milk removal in diary cows.

Biochemical studies showed that the mammary OT receptor is a 65-kDa protein (in rats), coupled to phosphoinositide turnover, and is most likely the same as the uterine-type OT receptor. In rats, Soloff and Wieder reported an ~100-fold increase in the number of OT receptors per mammary gland between the first day of pregnancy and late lactation. A very strong increase in OT binding sites toward term was also observed in porcine mammary tissues. In humans, however, Kimura et al. did not find an elevated expression level of the OT receptor mRNA during lactation. As judged by OT receptor immunoreactivity, unexpectedly, the ductal/glandular epithelium revealed a higher density of OT receptors compared with myoepithelial cells in both the lactating and nonlactating breast in humans and in the common marmoset (*Callithrix jacchus*).

Breast Cancer and Tumor Cells 2

Breast cancer is the leading cause of death in women between ages 35 and 45, but it is most common in women over age 50. It is unknown why mothers who breast-fed their babies have a 20% lower incidence of breast cancer after menopause than mothers who did not. Murrell proposed a hypothesis that partially links breast cancer to the activation of the OT system. Accordingly, carcinogens in the breast may be generated by the action of superoxide free radicals released when acinal gland distension causes microvessel ischemia. Thus inadequate nipple care in the at-risk years could lead to ductal obstruction preventing the elimination of carcinogens from the breast. The regular production of OT from nipple stimulation would cause contraction of the myoepithelial cells, relieving acinal gland distension and aiding the active elimination of carcinogenic fluid from the breast. Thus OT production was suggested as a preventative factor in the development of breast cancer both pre- and postmenopausally.

OT receptors have been described in a number of human breast tumours and breast cell lines, e.g., the human tumor cell lines MCF-7 or Hs578T. Copland et al. recently studied the behaviour of OT receptors in dependence of culture conditions in Hs578 cells. Hs578 cells responded to OT by an increase of intracellular $Ca^{2+}$and stimulation of ERK-2 phosphorylation and PGE_2synthesis. OT receptors in these cells were strongly downregulated by serum starvation. Conversely, restoration of serum and addition of dexamethasone (1 ¼M) increased the OT receptor levels by nearly 10-fold and also led to an elevation of the steady-state OT receptor mRNA level. These cells may be good models for future studies with respect to OT receptor regulation in dependence of steroids.

Controversial results have been published concerning a proliferative action of OT on breast tumor cells. Ito et al. found that OT had no effect on the growth of different breast cell lines cultured for 7 days. The OT receptor was expressed in breast cancer derived not from the myoepithelium but from the glandular or ductal epithelium. According to Sapino et al., OT exerts a trophic effect on myoepithelial cells in the mammary gland. In organotypic cultures of the mouse mammary gland, OT induced differentiation and proliferation of myoepithelial and, to a lesser extent, luminal epithelial cells. On the other hand, OT inhibited cell proliferation and tumor growth of rat and mouse mammary carcinomas. In MCF7 cells for instance, OT inhibited estrogen-induced cell growth and enhanced the inhibitory effect of tamoxifen on cell proliferation. Antiproliferative effects of OT have also been observed in various breast carcinoma cell lines as well as in human neuroblastoma and astrocytoma cells, and these effects were accompanied by the activation of the cAMP/protein kinase A pathway. In vivo, OT reduced the growth of certain mammary tumours. With the use of immunohistochemistry and RT-PCR, OT receptors and its corresponding mRNA were detected in normal and pathological breast tissues. Interestingly, the expression of OT receptors was positively correlated with the progesterone level of these tissues. Further studies are required to clarify the role of the OT system concerning the long-term effects in mammary tissues and tumor cell lines derived from them.

Kidney D

The kidney is one of the peripheral target tissues for neurohypophysial hormones that exercise control over hydromineral excretion. The hormones are released into the blood by stimulations, such as hypovolemia or hyperosmolarity, that generally activate the OT and AVP neurons. When plasma sodium concentration exceeds 130 mM, the levels of both hormones increase as an exponential function of plasma sodium concentration. OT is a nonhypertensive natriuretic agent. It is involved in normal osmolar regulation, which is presumably different from the volume regulatory components of Na^+ homeostasis. Acute administration of OT to conscious rats produced a modest increase in the glomerular filtration rate and effective filtration fraction. The natriuretic effect of OT is mainly due to a reduction in tubular Na^+ reabsorption, probably in the terminal distal tubule or the collecting duct.

Autoradiographical analysis showed the existence and precise localisation of OT receptors in the rat kidney. Interestingly, the distribution of OT binding sites undergoes reshaping during postnatal development as it was similarly observed in the rat brain during maturation. Specific OT binding sites have been first detected at embryonic *day 17* in the cortex. In the medulla, OT binding sites were first detected at embryonic*day 19* when this region is forming. In the adult rat, OT binding sites were exclusively localised in the cortex, mainly concentrated in the macula densa. In the inner medulla, the

OT binding sites were found on the loops of Henle of the juxtamedullary nephrons. Interestingly, at all stages examined, cortical OT-binding sites had a higher selectivity for OT versus vasopressin compared with the medullary sites. It was therefore suggested that OT binding sites of the macula densa and thin Henle's loop could represent two subtypes of OT receptors. The location profile of the receptors suggests a possible role for OT in the regulation of tubuloglomerular feedback and solute transport.

Estrogen induced OT receptor gene expression in the outer medulla region and increased expression of OT receptors in macula densa cells of ovariectomised female and adrenalectomised male rats. Experiments with the antiestrogen tamoxifen suggested cell-specific regulation of OT receptor expression in macula densa and proximal tubule cells. Thus OT receptors may mediate estrogen-induced alterations in renal fluid dynamics. OT receptor mRNA levels have been measured in kidneys of late-pregnant, peri-parturient, and lactating rats. Of interest, OT receptor transcripts could not be detected in renal tissues of peri-parturient females, at times when OT receptor mRNA levels were highest in uterus. However, OT receptor gene expression in macula densa cells gradually reappeared and again achieved control levels by *day 20* post partum. Breton et al. observed that both the OT receptor mRNA levels as well as the OT binding capacity of rat renal extracts were significantly lower at term. Although in the rat kidney there was no indication for another than the uterine-type OT receptor gene expression, the mechanisms controlling the expression of OT receptor genes in uterus, pituitary, and kidney are obviously different.

OT injected intraperitoneally caused dose-dependent increases in urinary osmolality, natriuresis, and kaliuresis in the rat. Moreover, the injection of OT evoked concomitant release of ANP and natriuresis. Soares et al. recently provided evidence for the hypothesis that OT acts in concert with ANP to induce natriuresis and kaliuresis via the common mediator cGMP. In rats, the OT-induced natriuresis was caused mainly by decreased tubular Na^+reabsorption and was accompanied by increases in both urinary cGMP and NO_3^- excretion. Although the natriuretic action of ANP is mediated by activation of renal guanylyl cyclase A (GC_A) receptors localised in glomeruli, their afferent and efferent arterioles, and the tubules, OT is hypothesized to induce natriuresis by activation of renal NO synthase localised in macula densa cells. Presumably, binding of OT to their receptors on NOergic cells in the proximal tubules and macula densa leads to an increase of intracellular [Ca^{2+}], followed by Ca^{2+}/calmodulin-induced stimulation of NO synthase and activation of the soluble guanylate cyclase by NO. The consequent release of cGMP could then mediate natriuresis and kaliuresis via closure of Na^+ and K^+ channels. In cortical collecting duct cells, it has been shown that the NO-inhibited Na^+ transport is associated with increased cGMP content. Overall in rats, OT as well as ANP induce a concomitant reduction of both extracellular and intracellular fluid volume in states of increased body fluid volume.

The signal transduction of the renal OT receptor has been evaluated in various kidney epithelial cells in culture, e.g., LLC-PK_1. In these cells, OT induced stimulated phosphoinositide hydrolysis and transiently increased cytosolic [Ca^{2+}]. OT was also able to stimulate the soluble guanylate cyclase and increased the intracellular level of cGMP in LLC-PK_1 cells. Moreover, OT was found to stimulate the synthesis of prostaglandins in a dose-dependent manner in kidney homogenates. What is known about the OT effects on kidney function in species other than rats? In rabbits, OT was shown to affect apical sodium conductance of the microperfused cortical collecting duct through OT receptors distinct from the vasopressin receptors. The effects were inhibited by the addition of an OT antagonist. Although the natriuretic response to OT has also been described for conscious dogs, it probably does not occur in humans and nonhuman primates. Thus the contribution of OT to renal physiology in primates including humans, if any, remains uncertain.

Heart and Cardiovascular System E

Peripherally injected OT decreases mean arterial pressure in rats. Even in the absence of a central control mechanism, OT is able to reduce the heart rate and the force of atrial contractions in isolated atria from perfused rat hearts. An OT antagonist reversed the bradycardia caused by OT. Moreover, administration of OT at high concentrations (1 ¼M) leads to the stimulation of ANP release. An OT antagonist first inhibited this ANP release, and after prolonged perfusion, it finally decreased the OT-induced ANP release below that of control hearts. This suggested that intracardial OT stimulates ANP release in the heart. Gutkowska et al. proposed that OT and ANP act in concert in the control of body fluid and in cardiovascular homeostasis. In favour of this hypothesis, OT receptor transcripts and OT binding sites were shown to be present on atrial and ventricular sections as detected by in situ hybridisation and autoradiograpy, respectively. Furthermore, the OT receptor gene is expressed in all chambers of the rat heart, and the analysis of the RT-PCR products indicated the presence of the uterine-type OT receptors in the heart. Although the OT receptor mRNA levels in the atria were found to be higher than in the ventricles, overall the OT receptor mRNA levels were calculated to be at least 10 times lower than the OT receptor mRNA level present in the uterus of a nonpregnant rat.

The rat heart was also shown to be a site of OT synthesis. OT was detected in the effluent of isolated heart perfusates as well as in the medium of cultured atrial myocytes. OT concentrations were found to be higher in the atria than in the ventricles. In the right atrium, OT concentrations ~20-fold higher than those in the rat uterus have been measured. In contrast, the OT mRNA level in the heart tissues was found to be lower than that in the rat uterus. This discrepancy argues against the postulated abundant biosynthesis of cardiac OT. The relatively low concentrations of OT in the heart chambers and the high OT doses required for ANP release would be more compatible with paracrine or autocrine effects of cardiac OT, particularly in the right atrium. Although the OT quantities released from the perfused heart are not sufficient to substantially change the plasma concentrations of OT, they may contribute to the natriuretic action via stimulation of ANP release. Presumably, blood volume expansion via baroreceptor input to the brain causes the release of OT that circulates to the heart. OT-induced ANP release in the heart may be achieved after activation of OT receptors and subsequent elevation of intracellular [Ca^{2+}], which in turn could stimulate exocytosis and ANP secretion. ANP then exerts a negative chrono- and inotropic effect via activation of guanylyl cyclase and release of cGMP. Finally, a rapid reduction in the effective circulating blood volume is produced by an acute reduction in cardiac output, coupled with ANP's peripheral vasodilating actions. The ANP released would also act on the kidneys to cause natriuresis, and ANP acts within the brain to inhibit water and salt intake, leading to a gradual recovery of circulating blood volume to normal. Since the plasma concentration of both OT and ANP were found to be increased after parturition, the OT-stimulated ANP release might be at least partly responsible for the massive diuresis observed postpartum.

The effects of repeated subcutaneous OT injections on blood pressure and heart rate were investigated in spontaneously hypertensive rats. Surprisingly, when OT were given for 5 days, a sustained decrease in blood pressure was observed in male but not female rats, whereas the heart rate was unaffected. Moreover, acute versus chronic OT treatments caused opposite effects on blood pressure, and these effects were modified by female sex hormones. It appears that the complete OT system is present in the vasculature of the rat. The OT concentrations in the aorta and vena cava were reported to be even higher than those in the right atrium of the heart. Thus OT might play a direct role in volume and pressure regulation in a paracrine/autocrine manner. In the case of the umbilical-placental vasculature, OT was reported to be a far more potent vasoconstrictor than vasopressin. OT was therefore proposed to be involved in the closure of the umbilical vessels at birth. A uterine-type OT receptor could also mediate vasodilatory responses in human vascular endothelical cells.

OT-induced cardiac effects were also reported for some other species. The cardiac effects of neurohypophysial hormones were studied in the frog, *Rana tigrina*, and in the snake, *Ptyas mucosa*. In both species, OT produced dose-related cardiac effects in isolated atrial preparations. In the frog, vasotocin was the most potent hormone, whereas in the snake, OT produced the most potent dose-related positive inotropic changes. Furthermore, OT was found to modulate the intrathoracic sympathetic ganglionic neurons regulating the canine heart. When OT was injected into right atrial ganglionated plexi, heart rate and atrial forces were reduced in 5 of 10 dogs studied. However, no cardiac changes occurred when OT was injected into left atrial or ventricular ganglionated plexi. In this study, the connectivity with the central nervous system (CNS) was found to be necessary to elicit consistent OT-induced responses. Although OT elicited neuronal and concomitant cardiovascular responses in most dogs when administered adjacent to spontaneously active intrinsic cardiac neurons, OT injection into the superior vena cava only evoked a slight systemic hypotension in two of seven dogs.

Other Localisations F

Thymus 1: The thymus is the primary lymphoid organ reponsible for the selection of the peripheral T-cell repertoire. Neuropeptides and their receptors have been described in the thymus, supporting the concept of a close dialogue between the neuroendocrine and the immune systems at the level of early T-cell differentiation. The neurohypophysial peptides have been shown to trigger thymocyte proliferation and could induce immune tolerance of this conserved neuroendocrine family. OT has been identified in human thymus by immunoreactivity and at the transcriptional level. OT was present in the thymus in surprisingly large amounts and was found to be restricted to certain subtypes of thymic epithelial cells. Due to its higher expression level in the thymic epithelium, OT was proposed as the self-antigen of the neurohypophysial family. OT was found to be colocalised with the cytokeratin network of thymic epithelial cells and not within secretory granules. The peptide is therefore not secreted but behaves like an antigen presented at the outer surface of the cell. Thymic OT also behaves as a cryptocrine signal targeted at the epithelium membrane from where it is able to interact with neurohypophysial peptide receptors expressed by pre-T cells. OT receptors are predominantly expressed by cytotoxic $CD8^+$lymphocytes, and they transduce signals via the phosphoinositide pathway. In pre-T cells, OT was found to induce the phosphorylation of focal adhesion kinase. Thus it was suggested that OT actively intervenes in the program of T-cell differentiation both as a neuroendocrine self-antigen and as a promoter of T-cell focal adhesion following a cryptocrine pathway. OT receptors detected in thymic membranes or on thymocytes revealed a ligand selectivity similar to that of uterine OT receptors. Caldwell et al. reported that sexual activity leads to decreases of OT receptor densities in the thymus of rats. In human thymus extracts, the content of the OT immunoreactivity declined with increasing age, whereas in rat thymic extracts, it was reported to increase during aging. The age-dependent changes might be linked to thymic involution. Some immune pathologies in humans may be explained by thymic OT being involved in T-cell-positive selection and activation.

Fat Cells 2: In adipocytes, OT has a so-called insulin-like activity in that it stimulates glucose oxidation and lipogenesis and increases pyruvate dehydrogenase activity. Treatment of rat adipocytes with lipolytic stimuli such as glucagon or isoprenaline increased the conversion of choline into phosphatidylcholine, and this lipolytic effect was antagonised by OT. In rat adipocytes, OT at a concentration of 1 nM markedly increased phosphoinositide breakdown. OT also elicited a transient increase in intracellular $[Ca^{2+}]$ and stimulated the protein kinase C activity. Moreover, human fat cells possess a plasma membrane-bound H_2O_2 generating system that is sensitive to extracellular stimuli. OT as well as insulin were able to activate this system. It was suggested that the H_2O_2 produced might participate in the regulation of fat cell differentiation and/or maintenance of the differentiated state. Populations of low- and high-affinity OT receptors have been described in rat adipocytes.

Pancreas 3: OT and AVP have been identified in human and rat pancreatic extracts at higher concentrations than those found in the peripheral plasma. However, a local synthesis of these peptides within this organ has not yet been established. According to most studies in several species, the neurohypophysial hormones induce the release of glucagon and, to a lesser extent, insulin from the pancreas. During in situ perfusion of the rat pancreas with OT, a marked stimulation of glucagon release and a modest stimulation of insulin release were observed. With the use of isolated islets from rat pancreas, OT elicited a stimulation of glucagon release but failed to influence insulin release in a culture medium with low glucose content. The glucagonotrophic action of OT was greatly diminished in the presence of a higher concentration of glucose. Similar findings were reported in studies with conscious dogs. With the use of the microdialysis technique, OT administered directly in the pancreas of the rat stimulates the release of both insulin and glucagon. OT binding sites were identified in the periphery of the islets of Langerhans, corresponding to the localisation of the glucagon-producing α-cells. In isolated mouse islets, OT produced an increase in somatostatin, glucagon, and insulin release, and it amplified the glucose-induced insulin release probably via stimulation of the phosphoinositide metabolism. In humans, OT evoked a rapid surge in plasma glucose and glucagon levels followed by a later increase in plasma insulin and epinephrine levels. The effects of OT on plasma glucagon and epinephrine levels were potentiated by hypoglycemia. OT was also found to potentiate glucose-induced insulin secretion. In contrast, Page et al. observed no effects of OT on the decline or recovery of blood glucose concentrations or on the plasma glucagon response to insulin-induced hypoglycemia in humans.

Impaired glucose tolerance and hyperinsulinemia are common features of obesity. Interestingly, the plasma levels of OT were found to be fourfold higher in male and female obese subjects compared with control subjects. OT rises in hypoglycemia, and this response is partially inhibited by dexamethasone. The OT rise in response to insulin-induced hypoglycemia was reduced in obese men. Pretreatment with the opioid antagonist naloxone enhanced the OT response to hypoglycemia in obese men and suggested an abnormal activity of endogenous opioids in obesity. In women, unlike men, endogenous opioids did not modulate OT release during insulin-induced hypoglycemia.

A pancreatic OT receptor was cloned from the rat insulinoma cell line RINm5F. CHO cells that have been transfected with this receptor responded to OT with an increase in cytosolic Ca^{2+} concentration and inositol phosphate levels as well as arachidonic acid release and

PGE_2 synthesis. Amino acid sequence homology, binding specificity, and the ligand-activated signal transduction pathways suggested that the pancreatic OT receptors expressed in RINm5F cells are indistinguishable from the uterine-type OT receptors that have been described in other tissues.

Adrenal Gland 4: Ang and Jenkins first identified immunoreactive OT and AVP in human and rat adrenal glands. OT predominanted over AVP, and both peptides occurred at concentrations far greater than those found in plasma. Immunohistochemical studies in adrenal glands from rat, cow, hamster, and guinea pig showed that OT was localised in both the cortex and medulla in all these species. In the cortex, the OT immunoreactivity was most intense in the zona glomerulosa, whereas in the medulla, the OT staining was higher but revealed more species variation. Bovine adrenal medulla membranes revealed high-affinity low-capacity OT binding sites. OT binding was also detected on primary monolayer cultures of bovine adrenal chromaffin cells. However, the capacity of these sites was near the detection limit, and OT concentrations in the micromolar range were necessary to stimulate the phosphoinositide pathway. From the few studies that have addressed the role of OT in the adrenal gland, no clear picture has emerged. Perfusion of the isolated rat adrenal gland with OT at 100 nM inhibited the acetylcholine-stimulated aldosterone secretion, whereas smaller doses of OT given as a bolus were able to stimulate aldosterone secretion in the intact perfused rat adrenal gland, but not in superfused adrenal cells. Legros et al. hypothesized that OT acts also at the adrenal gland level to decrease cortisol release and/or synthesis in humans. In addition, a proliferative effect of OT was suggested by the observation that the number of chromaffin cells in the adrenal medulla of rats was increased in response to OT.

The list of peripheral tissues containing OT receptors is certainly not complete. Recently, functional OT receptors have also been discovered in primary cultures of human osteoblasts and in a human epithelial osteosarcoma cell line (Saos-2). Further studies are required to show whether OT is a bone anabolic agent as suggested by the authors.

The Central Oxytocin System

Localisation Profile A

Localisation of OT 1: In the CNS, the OT gene is primarily expressed in magnocellular neurons in the hypothalamic PVN and SON. Action potentials in these neurosecretory cells trigger the release of OT from their axon terminals in the neurohypophysis. In the PVN, two populations of OT-staining neurons have been identified: magnocellular neurons that terminate in the neurohypophysis and parvocellular neurons that terminate elsewhere in the CNS. Only a small fraction (0.2%) of the OT neurons were calculated to possess axon collaterals to both the neurohypophysis and the extrahypothalamic areas. Although few OT perikarya were observed other than in the magnocellular nuclei, OT fibres and endings have been described in various brain areas in the rat: the dorsomedial hypothalamic nucleus, several thalamic nuclei, the dorsal and ventral hippocampus, subiculum, entorhinal cortex, medial and lateral septal nuclei, amygdala, olfactory bulbs, mesencephalic central gray nucleus, substantia nigra, locus coeruleus, raphe nucleus, the nucleus of the solitary tract, and the dorsal motor nucleus of the vagus nerve. OT fibres also run toward the pineal gland and the cerebellum, with most of them continuing toward the spinal cord. A few OT fibres end on the portal capillaries in the median eminence.

OT concentrations in the extracellular fluid of the SON were calculated to be >100- to 1,000 fold higher than the basal OT concentration in plasma, i.e., more than 1–10 nM. High-frequency electrical discharges of OT neurons as they occur, e.g., during the milk ejection reflex, might release even higher local OT concentrations.

Plasma OT does not readily cross the blood-brain barrier, and there is no relationship between the release of OT into the blood by the neurohypophysis and the variations in OT concentrations in the cerebrospinal fluid (CSF). Peripheral stimulations such as suckling or vaginal dilation that elicit large increases in plasma OT may or may not change the concentration of OT in the CSF. As shown in rats, electrical stimulation of the neurohypophysis only evokes the release of OT into the blood, whereas stimulation of the PVN elicits a release of OT into the blood and into the CSF. After hypophysectomy, OT disappears from the blood, whereas its concentration increases in the CSF. The OT in the CSF is probably derived from neurons that extend to the third ventricle, the limbic system, the brain stem, and the spinal cord. In the CSF, OT is normally present at concentrations of 10–50 pM, and its half-life is much longer (28 min.) than in the blood (1–2 min.). In humans and in monkeys, a circadian rhythm in the OT concentrations in the CSF has been found with peak values at midday. No such circadian rhythms have been observed in the CSF of rats, cats, guinea pigs, or goats. Circadian rhythms have never been observed in plasma OT concentrations.

Further complicating factors are the presence of an OT-like peptide and the appearance of conversion products of OT that possibly exert their effects exclusively on the brain, e.g., to influence learning and memory processes. Moreover, OT fragments such as OT-(1-6) or OT-(7-9) could cross the blood-brain barrier more easily.

Localisation of OT Receptors 2: Experiments with primary cell cultures showed that OT receptors are localised both on hypothalamic neurons and astrocytes. Concerning the regional distribution of OT binding sites in the brain, marked species differences have been observed. In rats, OT receptors are abundantly present in several brain regions, i.e., some cortical areas, the olfactory system, the basal ganglia, the limbic system, the thalamus, the hypothalamus, the brain stem, and the spinal cord. In the adult rat, a high density of OT receptors is found in the dorsal peduncular cortex, the anterior olfactory nucleus, the islands of Calleja and ventral pallidum cell groups, the limbic system (bed nucleus of the stria terminalis, central amygdaloid nucleus, ventral subiculum), and the hypothalamic ventromedial nucleus. OT receptor mRNA was detected in brain areas mostly coinciding with the occurrence of OT binding sites. OT receptors were detectable in all spinal segments, but in low amounts and restricted to the superficial layers of the dorsal horn. No major differences in receptor distribution were observed between male and female brains. Notably, the distribution pattern of OT binding sites is markedly different from that of binding sites for AVP. Whenever present in the same area, OT and AVP binding sites were located in different parts of the area.

The distribution and numbers of OT binding sites undergo major changes during development. As shown by Tribollet et al., only a fraction of OT receptors is constantly present during the development. Some OT receptors are only transiently expressed on neurons, e.g., during infancy or during maturation. In male and in female rats, two critical periods during the development were recognised: the third postnatal week, which precedes weaning, and puberty. For example, in the infant brain, the cingulate and retrosplenial cortex as well as the substantia gelatinosa of the spinal cord contained the highest densities of OT binding sites, whereas low or undetectable numbers of OT receptors were observed in these areas in the adult rat brain. Conversely, OT receptors in some other areas were abundant in the adult brain but undetectable before puberty, e.g., in the olfactory tubercle (Calleja islands and ventral pallidum cell groups) and in the hypothalamic ventromedial nucleus. During aging, however, the number of OT binding sites decreased again in the latter areas. This was probably mediated by the markedly lower level of plasma testosterone in aged rats. In fact, the expression of OT receptors in the olfactory tubercle and in the ventromedial hypothalamic nucleus was shown to be dependent on gonadal steroids, and testosterone treatment of aging rats could restore normal adult levels of OT receptors in these areas.

As already mentioned, there is a high diversity of OT receptor distribution between different species. For example, the ventral subiculum in the hippocampus contains high densities of OT binding sites in the rat, whereas in the guinea pig, the hamster, the rabbit, and the marmourset, no OT binding sites were detected in this area. In the rabbit, no OT receptors were found in the hypothalamic ventromedial nucleus. In monogamous versus polygamous voles, OT receptor distribution was shown to reflect social organisation. In human brains, dense OT receptor binding sites were visualised in the pars compacta of the substantia nigra unlike in several other species examined so far. Thus, in humans, nigrostriatal dopamine neurons could be a target for OT, and the OT system may be involved in motor and other basal ganglia-related functions. Strong OT binding intensity was also observed in the basal nucleus of Meynert, but OT binding sites were lacking in the hippocampus, amygdala, entorhinal cortex, and olfactory bulb of human brains.

In a few brain areas such as the ventral hippocampus and the bed nucleus of the stria terminalis (BNST), the reactivity of neurons to OT could be related with OT axon endings and OT receptors. The distribution of OT binding sites in the rat spinal cord was shown to coincide with that of the OT innervation, suggesting that OT is involved in sensory and autonomic functions. In most other brain areas, it was not possible to identify a clear relationship between the presence of the whole OT system and the physiological data. So, in male rats, Hosono et al. identified a functional OT receptor system in the subfornical organ, where in binding studies the expression is almost neglected. Marked receptor-peptide mismatches exist in some brain regions, for example, in the amygdala, where very few OT afferents but highest OT binding activities have been found. On the other hand, one has to recognise that OT binding sites in regions with high OT concentrations may be downregulated to a level that may not be detectable by autoradiography using radioligands. In the lactating rat, for instance, OT binding sites have been found to be nearly undetectable in den PVN and SON. As soon as 5–20 min. after intracerebroventricular injection of an OT antagonist, a strong increase of OT binding sites was noticed in the magnocellular nuclei. In this case, the presence of the OT antagonist may prevent the downregulation of OT receptors induced by the high concentrations of OT that are released into the magnocellular nuclei during lactation.

In the brain, estrogen has only a modest influence on the synthesis of OT, but it has a pronounced effect on the regulation of the OT receptor. OT receptors (but not AVP receptors) are regulated by gonadal steroids in the rat brain in a complex fashion. Castration and inhibition of aromatase activity reduced, whereas estradiol and testosterone increased OT binding, particularly in regions of the brain assumed to be involved in reproductive functions, such as the ventrolateral part of the hypothalamic ventromedial nucleus (VMN) and the islands of Calleja and neighbouring cell groups. Estrogen treatment increased the affinity of OT receptors in the medial preoptic area-anterior hypothalamus and increased both the density and the area of OT binding in the rat VMN. Progesterone caused a further increase in receptor binding and was required for a maximal extension of the area covered by OT receptors. In another study, chronic progesterone treatment increased basal OT receptor density in the limbic structures, decreased it in the ventromedial nucleus, and prevented estrogen-induced increases in ligand binding in all areas studied with the exception of the medial preoptic area. Glucocorticoids have also been reported to modulate cerebral OT receptors. The brain OT receptor varies across species not only in its distribution but also in its regional regulation by gonadal steroids. For example, estrogen increased OT receptor binding in the rat brain but reduced OT receptor binding in the homologous regions of the mouse brain.

Adenohypophysis C

It has been suggested that some hypothalamic OT reaches the anterior pituitary lobe via the hypothalamo-pituitary portal vasculature. OT might thus be able to influence anterior pituitary hormones as a hypothalamic regulating factor. OT is present in nerve terminals in the median eminence. It was found to be released into the portal vessels, and specific OT receptors are present in the rat adenohypopophysis. Another pathway for OT delivery to the adenohypophysis might be the short portal vessels connecting the posterior and anterior lobes. OT may participate in the physiological regulation of the adenohypophysial hormones prolactin, ACTH, and the gonadotropins.

There was a long controversy on whether OT released in response to suckling was responsible for the concomitant secretion of prolactin from the adenohypophysis. During suckling and under stress, both hormones are quantitatively predominant among the factors released. OT could only act as prolactin releasing factor when the dopamine levels are low, e.g., during the brief periods of dopamine withdrawal that characterise the onset of prolactin secretion under various physiological stimuli. With the use of a cell-specific targeting approach, it was shown that a subpopulation of lactotrophs respond to OT. Breton et al. demonstrated that the pituitary OT receptor gene expression is restricted to lactotrophs and dramatically increases at the end of gestation or after estrogen treatment. These findings question a direct function of OT on pituitary cells other than lactotrophs and suggest that OT might exert its full potential as a physiological prolactin-releasing factor only toward the end of gestation.

The major endocrine response to stress is via activation of the hypothalamic-pituitary-adrenal axis. ACTH secretion from the anterior pituitary is primarily regulated by CRH and AVP synthesized in neurons of the PVN. Unlike ACTH, plasma OT does not increase in response to all kinds of stress. In rats, OT can potentiate the release of ACTH induced by CRH. OT was also reported to stimulate ACTH secretion from corticotrophs in fetal and adult sheep. CRH is responsible for the immediate secretion of ACTH in response to stress. However, when the CRH levels are decreased following prolonged stress, the persistent level of OT in the median eminence could become important for the delayed ACTH response and for the generation of pulsatile ACTH secretory bursts.

In contrast, OT infusion into human volunteers actually inhibited the plasma ACTH responses to CRH. Suckling and breast stimulation in humans produced an increase in plasma OT and a decrease in plasma ACTH level. The observed negative correlation of both hormones indicates an inhibitory influence of OT on ACTH/cortisol secretion under a certain physiological condition in humans. Conclusively, OT might control ACTH release under some physiological conditions in a species-specific manner.

Gonadotropes in the adenohypophysis synthesize and secrete the two gonadotropin hormones, luteinizing hormone (LH) and follicle-stimulating hormone (FSH). Although gonadotropin-releasing hormone (GnRH) is believed to be the primary secretagogue for LH, OT has also been shown to stimulate LH release. OT administered to proestrous rats caused advancement of the LH surge and earlier ovulation. In addition, OT antagonists inhibited the peak of LH at proestrous. On the other hand, OT receptors have not been detected on gonadotropes but were identified on a gonadotrope-derived cell line. Indirect effects of OT on LH release have also been discussed. OT has been observed to synergistically enhance GnRH-stimulated LH release. OT may sensitise the pituitary before full GnRH stimulation. In human females, preovulatory OT administration promoted the onset of the mid-cycle LH surge. Overall, the physiological connection between OT and LH release has yet to be definitively established.

Centrally Mediated Autonomic and Somatic Effects D.

Cardiovascular Regulation 1: In the rat, OT-containing axons terminate in several brain stem nuclei known to be involved in cardiovascular control. High concentrations of OT are found in the nucleus of the solitary tract, which receives sensory inputs from the viscera, including baroreceptors of the cardiovascular system. Nevertheless, microinjection of OT into the nucleus tractus solitarius did not change the mean arterial pressure or the heart rate in the rat. In contrast, injection of OT into the dorsal motor nucleus of the vagus reduced the heart rate, and this effect was eliminated by an OT antagonist. In humans and rats, the bolus intravenous administration of OT is often associated with a decrease in blood pressure.

A role of central OT with respect to cardiovascular control has been established in many studies. Central pretreatment of rats with an OT antisense oligodeoxynucleotide attenuated the mean arterial pressure and heart rate responses induced by substance P. This suggests that OT neurons in the PVN mediate the increases in blood pressure and heart rate induced by stimulation of substance P receptors in the forebrain. An OT antisense oligonucleotide injected into the PVN abolished the tachycardia produced by shaker stress in rats, indicating that OT may act as mediator of stress-induced tachycardia. In addition, neuropeptide Y and peptide YY, which are central and peripheral modulators of cardiovascular and neuroendocrine functions, triggered OT release from perfused isolated terminals of the rat neurohypophysis. There is further evidence for an interaction between the central oxytocinergic and vasopressinergic systems in cardiovascular control. So, central injection of AVP (1–10 pmol) increased the mean arterial pressure and heart rate, and both responses were found to be enhanced in rats pretreated with OT. Recent findings suggested a putative interaction of OT with the ANPergic system.

Although atrial stretch releases ANP from cardiac myocytes, the response to acute blood volume expansion was found to be markedly reduced after elimination of neural control. Volume expansion distends baroreceptors in the right atria, carotid-aortic sinuses, and kidney and in turn alters afferent input to the brain stem and the hypothalamus. Because axons of the ANP neurons contained only low amounts of ANP (1,000 times less than in right atrium of the heart), it was hypothesized that hypothalamic ANP neurons cause release of OT, which circulates to the right atrium to stimulate ANP release. Because ANP has a negative ino- and chronotropic effect in the atrium, this would produce an acute reduction in cardiac output and, coupled with peripheral vasodilating actions, causes a rapid reduction in effective circulating blood volume.

Analgesia 2: Analgesic effects of OT have been reported in most but not all studies in mice, rats, dogs, and humans. Certain stimulations such as vaginal dilation led not only to a rise in plasma OT concentrations but also to an increase of the pain threshold. Dogs with neck and back pain caused by spinal cord compression had significantly more OT in their CSF than clinically normal dogs. Analgesia was observed in rats after injecting OT into the lateral ventricles. On the other hand, systemic OT did not produce analgesia in rats. Thus it was concluded that the observed increase in response latency in the hot-plate test may result from the sedative and vasoconstrictive effects of OT. Because the OT antagonist did not alter response latency on the hot-plate test, it seems unclear whether endogenous OT exerts a tonic effect on the pain threshold in rats.

In humans, intrathecal injection of OT was effective in treating low back pain for up to 5 h. An OT antagonist and the opiate receptor-blocker naloxone could reverse OT-induced analgesia. OT also increased β-endorphin and L-encephalin contents in the spinal cord, whereas an OT antagonist caused a decrease in the concentration of these opioids. Moreover, OT levels were elevated in the CSF of patients with chronic low back pain, perhaps a compensatory response to the painful condition. In a clinical case study, a high concentration (300 ¼g) of OT injected intravenously was reported to evoke strong analgesic effects lasting more than 70 min. in a patient with intractable cancer pain at a time when opiates were no longer effective.

Motor Activity 3: Centrally administered OT can induce or modify several forms of behaviour together with the associated motor sequences. OT increased general motor activity, and OT antisera decreased this hyperactivity and seizures in a complementary fashion. In this context, OT may possibly act at the spinal level. OT changed the spontaneous motor activity in female rats in strong dependence on steroid hormones. Treatment with low OT doses led to a decrease in peripheral locomotor activity, whereas increasing doses of OT provoked sedative effects as indicated by a suppression of locomotor activity and rearing. A maximal effect was obtained within 1 h and thereafter, the behaviour gradually returned to normal within 24 h. During static muscle contraction, blood pressure and heart rate reflexly increase as shown in anaesthetised cats. In this respect, OT may participate to regulate cardiovascular responses elicited by contraction of skeletal muscle.

Thermoregulation 4: In contrast to centrally administered AVP, for which antipyretic actions have been well documented, OT evoked, if at all, mostly weak antipyretic effects at higher concentrations. In rabbits, intracerebroventricular administered OT produced small but long-lasting hyperthermias that did not exceed 0.4°C. Intracisternally injected OT to adult male mice significantly increased colonic temperatures, with the maximal rise in temperature occurring 30 min. after administration of the peptide. OT also antagonised the hypothermia produced by other peptides such as bombesin or neurotensin. Conversely, whenever the body temperature was elevated in urethan-anaesthetised rats, e.g., in response to prostaglandin fever, OT was released into the ventral septal area, but not into the hippocampus. OT could also modulate the antipyretic action of AVP in rats. Possibly, the effects of OT with respect to thermoregulation may be physiologically significant during parturition and lactation.

Gastric Motility 5: In conjunction with the natriuretic effects of circulating OT, Verbalis et al. suggested the concept of a coordinated central and peripheral OT secretion as a mechanism for regulating body solute homeostasis in rats. The following findings support this concept. OT inhibits food intake in fasted rats and NaCl intake in hypovolemic rats. Inhibition of food and/or NaCl intake stimulates expression of c-*fos* in parvocellular and magnocellular OT neurons and indicates simultaneous activation of centrally projecting and pituitary-projecting OT neurons. Selective inactivation of brain cells containing OT-receptive elements leads to a disinhibition of NaCl intake similar to that produced by OT antagonists. In particular, gastric motility is inhibited by microinjection of OT into the dorsal motor nucleus or by intracerebroventricular administration of OT in rats. An OT antagonist administered intracerebroventricularly alone caused an increase in baseline gastric motility. In addition, stimulation of gastric vagal afferents by systemic administration of cholecystokinin (CCK) inhibited gastric motility, reduced food intake, and stimulated pituitary secretion of OT in rats. McCann and Rogers provided evidence that central OTergic neurons influence gastric motility and secretion by increasing the excitability of brain stem vagal neurons, i.e., sensory neurons in the nucleus tractus solitarius and motor neurons in the dorsal motor nucleus, which are both related to gastric function. According to Esplugues et al., the inhibitory effect of OT on the distension-stimulated gastric acid secretion in the rat perfused stomach is mediated by a nervous reflex involving a neuronal pathway that includes NO synthesis in the dorsal motor nucleus of the vagus.

For OT to have physiological relevance for maintenance of solute homeostasis, a coactivation of magnocellular and parvocellular OT ergic systems has been postulated: central, i.e., parvocellular OT as inhibitor of solute ingestion, and peripheral, i.e., pituitary secreted OT to enhance solute excretion via renal sodium excretion. However, pituitary OT secretion from magnocellular neurons is not always linked to decreased gastric motility in rats. For example, nipple attachment and sucking by pups, a well-known stimulus for neurohypophysial secretion of OT, did not decrease gastric motility in lactating rats. In view of the fact that OT promotes reproductive behaviours, it appears quite plausible that OT simultaneously acts as inhibitor for food intake, since both behavioural responses would conflict with each other. However, some findings do not fully support this concept or need to be clarified. First, central administration of an OT antagonist did not block the effects of CCK and LiCl to inhibit gastric motility.

This suggests that the parvocellular OT projections modulate the activity of intrinsic brain stem reflex arcs and do not exert a direct control over vagal efferent outputs. Second, OT receptors and mRNA localisations have been reported to occur predominantly on vagal neurons of the dorsal motor nucleus, which are generally excitatory to gastric motility, whereas OT is known to inhibit motility via neurons of the nucleus tractus solitarius. Finally, marked interspecies differences have been reported with respect to OT's relevance for maintenance of solute homeostasis. Whereas in rats nausea-producing agents and CCK each stimulated OT rather than AVP release, in humans and monkeys vasopressin but not OT secretion was provoked in response to these agents. In humans, nausea is a powerful stimulator of VP release induced by centrally acting emetics such as apomorphine, whereas OT is only weakly stimulated by apomorphine. In the rat, the converse is true. OT is stimulated by emetic agents, and VP shows little change. Therefore, we have little evidence that OT is involved in the physiology of emesis in humans. Intracisternal injection of OT also failed to induce emesis in dogs.

Hydromineral Regulation and Osmoregulation 6

In rats, the concerted release of OT and AVP plays a key role in osmoregulation through the effects of natriuresis and diuresis. Osmoreceptors are located in systemic viscera and in central structures that lack the blood-brain barrier. The nucleus tractus solitarius and lateral parabrachial nucleus receive neural signals from baroreceptors and are responsible for inhibiting the ingestion of fluids under conditions of increased volume and pressure and for stimulating thirst under conditions of hypovolemia and hypotension. Osmosensitive neurons combine with endogenously generated osmoreceptor potentials to modulate the firing rate of magnocellular neurons. In magnocellular neurons isolated from the SON of the adult rat, stretch-inactivated cationic channels transduce osmotically evoked changes in cell volume into functionally relevant changes in membrane potential. As judged from the distribution of Fos immunoreactivity, a fraction of oxytocinergic neurons in the subnuclei of the PVN were found to be activated during the satiety process of sodium appetite. Blackburn et al. provided evidence that salt appetite is regulated by both sodium- and osmolality-sensing mechanisms in rats. Using a selective receptor inactivation approach, these authors concluded that in contrast to the central ANP receptors that mediate both sodium- and osmolality-induced inhibition of NaCl ingestion, the central OT receptors primarily mediate osmolality-induced inhibition of NaCl ingestion.

In anaesthetised rats, electrical stimulation of renal afferents provided evidence that sensory information originating in renal receptors changes the activity of oxytocinergic neurons in the PVN. Thus renal receptors could contribute to the hypothalamic control of AVP and OT release into the circulation. In humans, there is neither an OT response to changes in plasma osmolality nor is there a natriuretic response to large doses of OT.

Central Behavioural Effects

Sexual Behaviour A

The "classical" peripheral target tissues for OT, uterus and mammary glands, are both organs linked to reproduction. There is also evidence that OT homologs in fish, amphibians, and reptiles as well as in molluscs and annelids are important in the control of reproductive behaviours. In the mollusc *Lymnaea stagnalis*, the conopressin gene encoding the putative ancestral receptor to the VP/OT receptor family is expressed in neurons that control male sexual behaviour, and its gene products are present in the penis nerve and the vas deferens. Thus a phylogenetically old connection may exist between systems controlling reproductive hormone release and reproductive behaviours. Moreover, the neurohypophysial release of OT into the circulation is most efficiently provoked by various kinds of stimulations of genitals and the breast in different mammalian species.

Unlike humans, the majority of animals breed only during certain times of the year, and the expression of sexual behaviour is under strict endocrine regulation. Male and female animals mate only when circulating steroid hormone levels are at appropriate concentrations, and to influence behaviour, the steroid hormones must profoundly affect the neurotransmission in the brain. In contrast, humans and other primates display a phenomenon called "concealed ovulation" that may have played a role in the evolution of social structures.

Female Sexual Behaviour in Animals 1

In the female rat, the sexual behaviour is generally divided into proceptive (soliciting) and receptive (lordosis) responses, both of which are under the control of gonadal steroids. Many studies have shown that intracerebral infusion of OT into the hypothalamus facilitates both aspects of the sexual behaviour in estrogen-primed female rats. In many but not in all studies, progesterone administration was necessary for the OT-induced facilitation of lordosis behaviour. In particular, prolonged (3 days) priming with estradiol induced lordosis, even in the absence of progesterone. Furthermore, the frequency and/or duration of lordosis was differentially affected, dependent on the site of OT administration. It was therefore suggested that OT may control different aspects of lordosis. Surprisingly, low concentrations of OT could even suppress lordosis when infused into the lateral ventricle of female rats.

On the other hand, several experiments with OT antagonists confirmed that endogenous OT is essential for the expression of sexual behaviour. So, infusion of an OT antagonist into the medial preoptic area before progesterone inhibited sexual receptivity and increased rejection in female rats. Additionally, in nonovariectomised female rats in spontaneous behavioural estrus, intracerebroventricular injection of OT increased lordosis quotient and lordosis duration, whereas an OT antagonist prevented this OT-induced effect. Witt and Insel observed that significant OT antagonist effects were absent in females primed with estradiol alone, suggesting that the OT antagonist attenuates progesterone facilitation of female sexual behaviour. Surprisingly, these behavioural effects of OT antagonist administration did not appear immediately, but emerged only when antagonist was given with progesterone 4–6 h before behavioural testing. Two other observations support further evidence for the involvement of endogenous OT in the expression of female sexual behaviour. First, the immunoreactivity of the immediate early gene product Fos was induced in OT neurons after sexual activity in female rats. Second, intrahypothalamic infusions of antisense oligonucleotides directed against the OT receptor mRNA reduced lordosis behaviour as well as OTR in the VMN of estrogen-primed rats.

How could OT act to induce the expression of lordosis? Vincent and Etgen showed that steroid priming promotes an OT-induced norepinephrine release in the ventromedial hypothalamus of female rats. This occurs most probably by a peripheral mechanism, e.g., vaginocervical contraction, leading to increased excitability of ventromedial hypothalamus neuronal activity and the expression of lordosis. With respect to OT effects on sexual behaviour, only few studies were performed in species other than rats. In ovariectomised hamsters (*Mesocricetus auratus*) primed with estradiol, microinjection of OT into the medial preoptic area-anterior hypothalamus or the ventromedial hypothalamus induced sexual receptivity, whereas injection of a selective OT antagonist reduced the levels of sexual receptivity. OT injection into the hypothalamus also increased ultrasonic vocalisations that female hamsters use to alert and attract potential mates.

Male Sexual Behaviour 2

Male sexual behaviours and performances have also been mostly studied in rats and are measured, e.g., by changes in the frequency or durations of mounts, penile erections, intromissions, ejaculations, and yawnings. Yawning is considered to be a phylogenetically old, stereotyped event that occurs alone or associated with stretching and/or penile erection under different conditions.

OT was found to be one of the most potent agents to induce penile erection in rat, rabbits, and monkeys. Melis et al. reported that low doses of OT (e”3 ng) can elicit penile erection and yawning in rats following site-specific injection into the PVN, i.e., the nucleus which contains cell bodies of the majority of oxytocinergic neurons projecting to extrahypothalamic brain areas. Both OT effects were prevented by intracerebroventricular injection of OT antagonists as well as by electric lesions of the PVN. According to Argiolas and Melis, the oxytocinergic neurons facilitate yawning by releasing OT at sites distant from the PVN, i.e., the hippocampus, the pons, and/or the medulla oblongata when activated by the D1/D2 dopamine agonist apomorphine, *N*-methyl-D-aspartate, or OT itself. This activation was antagonised by opioid peptides that prevent the yawning response. Central administration of OT also significantly reduced the latency to achieve ejaculation and shortened the postejaculatory interval. This effect was more pronounced in aged male rats and might be related to decreases in circulating testosterone. Apomorphine- and OT- induced penile erection and yawning were endocrine dependent and are differentially modulated by sexual steroids.

It was hypothesized that OT induces sexual behavioural responses by increasing NO synthase activity in the cell bodies of oxytocinergic neurons projecting to extrahypothalamic brain areas, e.g., via intracellular [Ca^{2+}] increase and stimulation of the Ca^{2+}/calmodulin-dependent NO synthase. In agreement with this suggested mechanism, OT-induced penile erections were prevented by PVN injections of É-conotoxin, a blocker of N-type voltage-dependent calcium channels. The PVN is one of the richest brain areas containing NO synthase, and NO is a well-known mediator of penile erection, both peripherally and centrally. Accordingly, injections of NO donors into the PVN induced penile erections, and this response was prevented by the intracerebroventricular administration of an OT antagonist. Conclusively, OT may induce male sexual behaviours by activating NO synthase in the PVN. The concomitant production of NO could cause the activation of oxytocinergic neurons projecting to extrahypothalamic brain areas such as the hippocampus and/or spinal cord and control the male sexual function.

However, not all studies are in favour of a central facilitatory effect of OT on male sexual activity. OT produced an immediate cessation in sexual activity of male prairie voles and remained so for at least 24 h. In another study, centrally injected OT prolonged the postejaculatory refractory periods. These and some other findings are consistent with the hypothesis that OT could also play a role in sexual satiety. OT-immunoreactive neurons in the parvocellular subnuclei of the PVN project to a sexually dimorphic motor nucleus in the lower spinal cord which innervates the striated bulbocavernosus muscle. This muscle is responsible for penile reflexes, and it was suggested that this pathway is involved in the integration of penile reflexes with other aspects of male copulatory behaviour that are under hypothalamic control.

Sexual Behaviour in Humans 3

It is well documented that levels of circulating OT increase during sexual stimulation and arousal and peak during orgasm in both men and women. Murphy et al. measured plasma OT and AVP concentrations in men during sexual arousal and ejaculation and found that plasma AVP but not OT significantly increased during arousal. However, at ejaculation, mean plasma OT rose about fivefold and fell back to basal concentrations within 30 min., while AVP had already returned to basal levels at the time of ejaculation and remained stable thereafter. Men who took the opioid antagonist naloxone before self-stimulation had reductions in both OT secretion and the degree of arousal and orgasm. A recent study confirmed that also in women peak levels of serum OT were measured at or shortly after orgasm. Carmichael et al. reported that the intensity of muscular contractions during orgasm in both men and women were highly correlated with OT plasma level. This suggests that some of OT's effects may be related to its ability to stimulate the contraction of smooth muscles in the genital-pelvic area. Enhanced sexual arousal and orgasm intensity was reported in a woman during intranasal administration of OT. This response could be elicited only while she was taking daily doses of an oral contraceptive with estrogenic and progestogenic actions and might be caused through direct effects on sexual organs or sensory nerve sensitivity. Overall, beyond its peripheral effects on reproductive organs, OT might affect or sensitise cerebral neurons responsible for the cognitive feelings of orgasm and could thus serve as a physiological substrate for both sexual behaviour and performances in humans as well as in animals.

Maternal Behaviour B

The development of maternal care has been well studied in rats. Nulliparous females display little interest in infants and when presented with foster young will either avoid or cannibalise them. At parturition, however, a dramatic shift in motivation occurs and maternal behaviours such as nest building and retrieval of pups became established. Pedersen and Prange first demonstrated that injection of OT into the lateral ventricles of nulliparous ovariectomised rats induces maternal behaviour. The rapid onset of maternal behaviour in response to OT has been confirmed in several studies. However, it is important to note that OT is effective only to initiate the maternal behaviour, but not for the performance of maternal behaviour per se. So, when the females become maternal or enter into estrus, an OT antagonist had no effect. Moreover, steroid priming was found to be essential for the initiation of maternal behaviours in all cases so far studied. Subcutaneous injection of the NO donor sodium nitroprusside was shown to prolong parturition and to inhibit maternal behaviour in rats. Because OT was able to restore intrapartum maternal behaviour, NO was suggested to interfere with the initiation of maternal behaviour via blocking or diminishing the release of OT. Genital stimulations that are known to activate the oxytocinergic system could also provoke maternal behaviours as shown in ewes and rats. Genital stimulations may also influence olfactory function through the activation of afferent noradrenergic pathways in the olfactory bulbs. The olfactory bulb is in fact regarded as a critical site where OT may induce onset of maternal behaviour. OT originating in the hypothalamic PVN possibly acts there to decrease olfactory processing. This is in line with earlier findings according to which anosmia favours the establishment of maternal behaviour. However, the female gets a variety of sensory stimulations from contact with the pups so that the neural circuits responsible for postpartum maternal behaviour are rather complex. In addition, recent studies demonstrate that the paternal genome has aquired the ability to regulate maternal behaviour by a epigenetic mechanism called genomic imprinting, which results in the parent-of-origin-specific expression of only one allele of a gene. A striking impairment of maternal behaviour was observed in mice in which neurally expressed imprinted genes of paternal origin were mutated or deficient. One hypothesis to explain this phenomenon is that the paternal interest is best served by prolonged care and feeding of his progeny through maternal lactation.

The role of OT for maternal behaviour became questionable by studies with transgenic mice, since female mice genetically deficient in OT displayed an apparently normal maternal behaviour. This clearly demonstrates that OT itself is not essential for maternal behaviours in mice. Of course, this does not rule out that other OT-like ligands are present to activate the OT receptors or that in other species than mice, OT is essential to develop normal maternal behaviours.

In humans, OT-related maternal behaviours have not been the subject of any systematic studies so far. In earlier reports it was shown that breast-feeding within 1 h of birth, when OT levels are very high, supports a long-lasting mother-infant bond and has a beneficial effect on the development of the child.

Social Behaviour C

Parental care, nursing, social interaction, pair bonding, and mutual defence are prototypical mammalian species behaviours and have been important for the successfull evolution of mammals. Love and social attachments function to facilitate reproduction, provide a sense of safety, and reduce anxiety or stress. These social behaviours are clearly opposed to the ancient self-preservative behaviours. When these balanced social interactions are disturbed, e.g., by a stressor, the self-preservative, fight-flight response pattern takes priority. Recent studies in rodents suggest that the neurohypophysial hormones in concert with steroids are key components in the central mediation of complex social behaviours, including affiliation, parental care, sexual behaviour, mate guarding, and territorial aggression. OT has been reported to mediate aggressive and affiliative behaviours in several species. Moreover, disruption of the OT gene in mice showed reduced aggression behaviours in some tests.

Social behaviours have been intensively studied in North American species within the genus *Microtus* (voles), which exhibit diverse forms of social organisation ranging from minimally parental and promiscous (e.g., montane vole, *M. montanus*) to biparental and monogamous (e.g., prairie vole, *M. ochrogaster*) social structures. The highly social prairie voles usually form long-term monogamous relationships as a consequence of mating and thereby develop a variety of affiliative behaviours, such as parental care, increased physical contacts with each other, and the attacking of adult intruders as a putative mate guarding response. Because vaginocervical stimulation concomitant with the intense mating of the prairie voles activates the oxytocinergic system, OT was suggested as a good candidate for the facilitation of pair bonding in prairie voles.

Moreover, in females that did not mate, intracerebroventricular but not peripheral administration of OT facilitated the formation of a pair bond. Conversely, intracerebroventricular injection of an OT antagonist before mating prevented partner preference formation in female prairie voles. This suggests that OT, released in response to mating behaviour, is sufficient and necessary for the female to form a pair bond to her mate. In males, however, AVP, but not OT, was shown to be critically involved in partner preference and selective aggression. In contrast, neither peptide appears to induce pair bonding in the promiscuos montane voles. Whereas OT had little effect on the social behaviour of montane voles, the same dose of AVP that induced aggression in the prairie vole increased self-grooming in the montane vole. Since the *Microtus*species share similar OT and AVP immunoreactive patterns but respond differently to exogenous peptide, the neuroendocrine differences between monogamous and polygamous species probably reside on the corresponding receptor molecules. In addition, the neuroanatomical distribution of OT and AVP receptors in several *Microtus* species suggested that the pattern of OT and AVP receptor binding is associated with social organisation.

High OT receptor densities were found in the prelimbic cortex and nucleus accumbens of prairie voles, but not montane voles. Because these brain areas are involved in the mesolimbic dopamine reward pathway, it was hypothesized that in the female prairie voles OT released in response to mating activates this reward pathway and may condition the female to the odour of her mate. Furthermore, in the montane vole, which shows little affiliative behaviour except during the postpartum period, brain OT receptor distribution changed within 24 h of parturition, concurrent with the onset of maternal behaviour. This suggests that variable expression of the OT receptor in brain could be an important mechanism in evolution of species-typical differences in social bonding and affiliative behaviour.

The complexity of the neuropeptide's actions on social behaviour is illustrated by studies in the squirrel monkey, in which the effects of exogeneous OT depended on the prior social status of individuals. Dominant males responded to OT administration with increased sexual and aggressive behaviour, whereas subordinate males displayed more associative behaviours. The differential behaviours might be due to the higher serum testosterone levels in dominant animals.

Transgenic mice created with the 52 -flanking region of the prairie vole OT receptor gene demonstrated that sequencing in this region influences the pattern of expression within the brain. Promoter sequences of the prairie vole OT receptor and V1a receptor genes and the resulting species-specific pattern of regional expression provide a potential molecular mechanism for the evolution of pair-bonding behaviours and a cellular basis for monogamy. Recently, Young et al. reported that an AVP-induced increase in affiliative behaviours could be observed in mice that were transgenic for the prairie vole V1a receptor gene.

Interactions among OT, AVP, and glucocorticoids could provide substrates for dynamic changes in social behaviours, including those required in the development and expression of monogamy. Results from research with voles suggest that the behaviours characteristics of monogamy may be modified by hormones during development and may be regulated by different mechanisms in males and females. Nonnoxious sensory stimulation associated with friendly social interaction induces a response pattern involving sedation, relaxation, and decreased sympathoadrenal activity. It is suggested that OT released from parvocellular neurons in the PVN in response to nonnoxious stimulation integrates this response pattern at the hypothalamic level. The health-promoting aspect of friendly and supportive relationships might be a consequence of repetitive exposure to nonnoxious sensory stimulation. On the other hand, OT and AVP could also be involved in the pathophysiology of clinical disorders, such as autism, which is characterised by an inability to form normal social attachments.

Stress-Related Behaviour D

In both male and female rats, OT exerts potent antistress effects such as decreases in blood pressure, corticosterone/cortisone level, and increases in insulin and CCK level. After repeated OT treatment, weight gain could be promoted, and the healing rate of wounds increased. Furthermore, acute exposure of rats to immobilisation stress resulted in an increase in OT mRNA level. OT secretion was also increased by forced swimming in virgin and early pregnant rats. In addition, microdialysis experiments on male rats showed specific increases in both central and plasma OT in response to forced swimming and shaker stress. Thus the stimulated release of OT could function to facilitate the activation of the hypothalamic-pituitary-adrenal axis and increase glucocorticoid release. In contrast to most other OT-induced behavioural and physiological effects, the antistress effects could not be blocked by OT antagonists, suggesting that yet unidentified OT receptors may exist.

It is also well documented that stress-induced central release of OT can ameliorate the stress-associated symptoms such as anxiety. So, OT has anxiolytic properties in estrogen-treated females in mice and in rats. Estrogen-induced increases in OT binding density in the lateral septum may contribute to the facilitation of social interactions. Like many other anxiolytic compounds, OT acts as an antidepressant as shown in two animal models of depression. Moreover, the antinociceptive action of OT is consistent with its antidepressant action. Some of these effects may be mediated by an influence of OT on dopaminergic neurotransmission in limbic brain regions.

Because stress and anxiety can impair maternal caretaking and reduce milk ejection, reduced stress responsiveness during lactation is adaptive for both mother and infant. Accordingly, lactating women had reduced hormonal responses to exercise stress when compared with postpartum women who bottle-feed their infants. In addition, women with panic disorder can experience a relief of symptoms during lactation. Because OT levels increase in blood and CSF after nonnoxious sensory stimulation such as touch, light pressure, and warm temperature in both female and male rats, OT could not only be responsible for the antistress effects occurring during lactation but may also be involved in health-promoting effects of relationships, social contacts, and networks.

Some of the OT-induced effects may be explainable by the anxiolytic actions of OT. For example, OT acting as an anxiolytic may reduce the inhibition inherent in social encounters. Also, behavioural tests in the laboratory frequently involve the exposure of the animal to a novel environment, combined with exposure to an unfamiliar conspecific. These stimuli are likely to induce a stress response, and perhaps this anxiety is reduced by OT. So, a unifying principal in OT action in the brain may be to facilitate social encounters by reducing the associated anxiety. However, more studies are necessary to clearly distinguish between physiological and pharmacological effects.

Feeding and Grooming E

OT acts as a "satiety hormone" in animals since both peripherally and centrally administered OT reduces feeding. In addition, food and anorexia-inducing agents, such as CCK, lead to pituitary OT secretion and subsequently to reduced food intake. This suggests that both nausea and satiety activate a common hypothalamic oxytocinergic pathway that controls the inhibition of digestion. Arletti et al. reported that OT, whether given intraperitoneally or intracerebroventricularly, reduced food consumption and the time spent eating, and it increased the latency to the first meal in fasted rats. Pretreatment with an OT antagonist completely prevented the feeding inhibitory effect of OT, and per se increased food intake. In particular, hyperosmolality is a very potent stimulus for OT release, and central OT was observed to mediate osmolality-related inhibition of salt appetite but was not essential for Na^+-sensing mechanism in salt appetite in rats. While

Arletti et al. reported that OT directly inhibited both food and water intake in rats, Olsen et al. observed a relatively small effect of OT on water intake when given in doses that significantly inhibited food intake. How could OT mediate its influence on the ingestive behaviours? PVN neurons in general, and OT projections in particular, could either act to modulate the activity of intrinsic brain stem reflex arcs or exert a direct control over vagal efferents that project to the gut and inhibit the gastric motility.

In mice, central administration of OT elicits dramatic behavioural excitation such as stress-induced escape, scratching, and grooming, particularly a pronounced self-grooming. Slow infusion of OT into the PVN also initiates self-grooming in rats. OT activates several aspects of grooming, preferentially genital grooming. The ability of OT to facilitate grooming involves activation of dopamine D1 receptor-mediated neurotransmission in the mesolimbic pathway as shown with knock-out mice lacking the D1 dopamine receptor. However, neurotransmitters other than dopamine (e.g., endorphins) may also play a supplementary role in neuropeptide-enhanced grooming in mice. Grooming is a behavioural response to stressors, and the induction of grooming might be part of a homeostatic mechanism for reducing anxiety and facilitation of pair bonds, whether between mother and infant or between mating partners, by increased transfer of odours between individuals.

Memory and Learning F

OT and AVP are considered to play a pivotal role in various aspects of learned behaviour. Overall, the concept emerged that AVP reinforces memory, whereas OT has just the opposite effect, namely, the attenuation of learning processes and memory. In particular, OT was shown to facilitate the extinction of avoidance reaction and to attenuate the storage of verbal memory. In cultured neurons, OT at concentrations over 1 M reduces the activity of NMDA receptors, thus impairing one of the major substrates for the induction of learning and memory. Regarding its impairing effects on some memory-related tasks, OT could possibly be involved in the forgetting of delivery pain in mothers. On the other hand, studies on rodents indicated that socially relevant behaviours are controlled in a more complex way by both AVP and OT released intracerebrally and argued against a simplistic view with respect to the memory-attenuating effects of OT. In this context, reports on the social memory in rats provide a good example. Social recognition can be measured in rats as the specific decrease in social investigation of a juvenile conspecies during the second encounter of the same individual. This social memory is based on the olfactory characteristics of the conspecies and is short lasting (<40 min.). OT had an impairing effect on the social memory in male rats. In another study, intraperitoneally injected OT to male rats significantly improved social memory. Depending on the dose, social memory was shown to be attenuated or facilitated by OT and derivatives as reported by Popik et al.. Interestingly, the active moieties for attenuation and facilitation of social memory reside in different parts of the OT molecule, e.g., the 5–7 and 8–9 region, respectively. However, the wide variety of observed effects has also led to the suggestion that OT has a more general effect on the cortical arousal rather than a specific effect limited to a certain stage of information processing.

Tolerance and Dependence to Opioids G

Drug tolerance, dependence, and addiction may involve neuroadaptive mechanisms related to learning and memory at cellular and systems levels. Chronic morphine administration alters the brain OT system, which suggests that OT might contribute to the behavioural, emotional, and neuroendocrine responses to opioids. Moreover, OT has been shown to inhibit the development of tolerance to morphine and to attenuate various symptoms of morphine withdrawal in mice. OT also decreased cocaine-induced locomotor hyperactivity and stereotyped behaviour in rodents. Dopaminergic neurotransmission and central OT receptors located in limbic and basal forebrain structures are probably responsible for mediating these various effects of OT in the opiate- and cocaine-addicted organism.

The influence of opioids on the secretory activity of OT neurons in the SON was studied in rats in more detail. The brain stem noradrenergic system that is activated, e.g., by peripheral administration of CCK, represents a major excitatory input to OT neurons. Morphine inhibits the OT neurons presynaptically via inhibition of the afferent noradrenergic input. When the opiate antagonist naloxone is administered locally in the SON, the inhibitory effect of morphine on CCK-stimulated norepinephrine release is reversed, and the firing rate of OT cells markedly increases. However, noradrenergic systems are not essential for the expression of morphine withdrawal excitation, since chronic neurotoxic destruction of the noradrenergic inputs to the hypothalamus did not affect the magnitude of withdrawal-induced OT secretion. Opioid receptors on OT neurons and/or on their afferent input may be activated by the endogeneous opioids pro-dynorphin and pro-enkephalin and are involved to restrain the secretory activity of OT cells. Interestingly, the efficiency of the endogeneous opioid restraint changes during pregnancy in rats. Therefore, Russell et al. have speculated that central opioid mechanisms could control OT neurons during parturition and can interrupt established parturition by inhibiting OT neuron-firing rate in disadvantageous environmental circumstances.

Central Disorders in Humans H

Obsessive-compulsive Disorder 1: Obsessive-compulsive disorder (OCD) includes cognitive and behavioural symptoms that are related to OT-associated behaviours. A possible role for OT in the neurobiology of a subtype of the OCD was suggested by the elevated CSF levels of OT and by the correlation between CSF OT levels and the severity of the disorder. Excessive grooming is often a component of OCD, and central OT administration induces high levels of self-grooming in animals. Periods of increased gonadal steroid secretion such as puberty and pregnancy are associated with a heightened risk for onset of OCD. Perhaps, gonadal steroids activate OT receptor during these periods. However, the results of another study do not support the hypothesis that OT might be a potential anticompulsive agent.

Eating Disorders: Anorexia and Prader-Willi Syndrome 2

The OT neurons of the PVN are good candidates for playing a physiological role in ingestive behaviour as "satiety neurons" in the human hypothalamus. In humans, OT appears not to be released in response to a meal. In women with anorexia nervosa, CSF levels of OT are reduced during the starvation phase of the illness and return to normal as the patients increase their food intake. Both hypoglycemia- and estrogen-induced OT increases were lower in underweight anorectic patients compared with normal controls. Anorectic subjects regained normal OT responsiveness to both stimuli after complete weight recovery. Reduced brain and plasma levels of OT during the starvation phase of the illness might contribute to the symptom profile of anxiety, loss of libido, amenorrhea, and increased activation of the hypothalamic-pituitary-adrenal hormonal stress axis.

In Prader-Willi syndrome, a genetic disorder characterised by gross obesity, insatiable hunger, hypotonia, hypogonadism, mental retardation, and a high risk for OCD, a 42% decrease in the number of hypothalamic OT neurons was observed. Moreover, there was a reduction in total cell number in the PVN, and the volume of the PVN-containing OT-expressing neurons was decreased by 54%. The lower number of OT neurons in the hypothalamic PVN is presumed to be the basis of the insatiable hunger and obesity of patients with the syndrome. Further evidence for hypothalamic and oxytocinergic dysfunction in Prader-Willi syndrome was provided by the finding that CSF OT was nearly twofold higher in patients with Prader-Willi syndrome compared with control subjects.

Depression and Schizophrenia 3

In postmortem samples from patients with depression, a 23% increase in the number of OT-immunoreactive neurons in the PVN of the hypothalamus was found. The putative higher rate of oxytocinergic activity in these patients is possibly associated with activation of the hypothalamic-pituitary-adrenal axis in these patients, since OT is known to potentiate the effects of CRH.

Increased concentrations of OT were observed in the CSF of schizophrenic patients, and they were higher in patients receiving neuroleptic treatment. However, Glovinsky et al. found no changes of the OT concentration in the CSF of schizophrenic patients either with or without neuroleptic medication. In another study, the basal level of OT-neurophysin was found to be threefold higher in the plasma of schizophrenics. So, impairments in the function of the OT system could play a role in the social deficits of schizophrenic patients, and OT is a candidate novel endogenous antipsychotic.

Neurodegenerative Diseases 4

OT injected into the hippocampus is reported to interfere with the formation of memory in experimental animals. In humans, high density of OT binding sites was observed in the nucleus basalis of Meynert, which is the major nucleus for cortical cholinergic neurons and degenerates in Alzheimer's disease. It has been reported that OT binding site density in the basal nuclei of Meynert decreases in the brains of patients suffering from Alzheimer-type senile dementia. OT concentration was found to be increased 33% in the hippocampus and temporal cortex of Alzheimer brains but was normal in all other regions examined. In two other sets of patients with Alzheimer dementia, the number of hypothalamic OT-expressing cells remained unaltered with aging. A slightly decreased number of OT-immunoreactive neurons was reported in the PVN of the hypothalamus in Parkinson's disease. However, more studies are needed to evaluate the role of OT for neurodegenerative disorders.

7

Fish Senses

Fish are a paraphyletic group of organisms that consist of all gill-bearing aquatic vertebrate (or craniate) animals that lack limbs with digits. Included in this definition are the living hagfish, lampreys, and cartilaginous and bony fish, as well as various extinct related groups. Most fish are ectothermic ("cold-blooded"), allowing their body temperatures to vary as ambient temperatures change, though some of the large active swimmers like white shark and tuna can hold a higher core temperature. Fish are abundant in most bodies of water. They can be found in nearly all aquatic environments, from high mountain streams (e.g., char and gudgeon) to the abyssal and even hadal depths of the deepestoceans (e.g., gulpers and anglerfish). At 32,000 species, fish exhibit greater species diversity than any other class of vertebrates. Fish, especially as food, are an important resource worldwide.

Commercial and subsistence fishers hunt fish in wild fisheries orfarm them in ponds or in cages in the ocean. They are also caught by recreational fishers, kept as pets, raised byfishkeepers, and exhibited in public aquaria. Fish have had a role in culture through the ages, serving as deities, religious symbols, and as the subjects of art, books and movies. Because the term "fish" is defined negatively, and excludes the tetrapods (i.e., the amphibians, reptiles, birds and mammals) which descend from within the same ancestry, it is paraphyletic, and is not considered a proper grouping in systematic biology. The traditional term pisces(also ichthyes) is considered a typological, but not a phylogenetic classification.

The earliest organisms that can be classified as fish were soft-bodied chordates that first appeared during the Cambrian period.

Although they lacked a true spine, they possessed notochords which allowed them to be more agile than their invertebrate counterparts. Fish would continue to evolve through the Paleozoic era, diversifying into a wide variety of forms. Many fish of the Paleozoic developed external armour that protected them from predators. The first fish with jaws appeared in the Silurian period, after which many (such as sharks) became formidable marine predators rather than just the prey of arthropods.

Diversity of Fish

Figure: *Fish come in many shapes and sizes. This is a sea dragon, a close relative of theseahorse. Their leaf-like appendages enable them to blend in with floating seaweed.*

The term "fish" most precisely describes any non-tetrapod craniate (i.e. an animal with a skull and in most cases a backbone) that has gillsthroughout life and whose limbs, if any, are in the shape of fins. Unlike groupings such as birds or mammals, fish are not a single clade but aparaphyletic collection of taxa, including hagfishes, lampreys, sharks and rays, ray-finned fish, coelacanths, and lungfish. Indeed, lungfish and coelacanths are closer relatives of tetrapods (such as mammals, birds, amphibians, etc.) than of other fish such as ray-finned fish or sharks, so the last common ancestor of all fish is also an ancestor to tetrapods. As paraphyletic groups are no longer recognised in modernsystematic biology, the use of the term "fish" as a biological group must be avoided.

Many types of aquatic animals commonly referred to as "fish" are not fish in the sense given above; examples include shellfish, cuttlefish, starfish, crayfish and jellyfish. In earlier times, even biologists did not make a distinction – sixteenth century natural historians classified alsoseals, whales, amphibians, crocodiles, even hippopotamuses, as well as a host of aquatic invertebrates, as fish. However, according the definition above, all mammals, including cetaceans like whales and dolphins, are not fish. In some contexts, especially in aquaculture, the true fish are referred to as finfish (or fin fish) to distinguish them from these other animals. A typical fish is ectothermic, has a streamlined body for rapid swimming, extracts oxygen from water using gills or uses an accessory breathing organ to breathe atmospheric oxygen, has two sets of paired fins, usually one or two (rarely three) dorsal fins, an anal fin, and a tail fin, has jaws, has skin that is usually covered with scales, and lays eggs.

Each criterion has exceptions. Tuna, swordfish, and some species of sharks show some warm-blooded adaptations—they can heat their bodies significantly above ambient water temperature. Streamlining and swimming performance varies from fish such as tuna, salmon, and jacks that can cover 10–20 body-lengths per second to species such as eels andrays that swim no more than 0.5 body-lengths per second. Many groups of freshwater fish extract oxygen from the air as well as from the water using a variety of different structures.Lungfish have paired lungs similar to those of tetrapods, gouramis have a structure called the labyrinth organ that performs a similar function, while many catfish, such as *Corydoras*extract oxygen via the intestine or stomach. Body shape and the arrangement of the fins is highly variable, covering such seemingly un-fishlike forms as seahorses, pufferfish,anglerfish, and gulpers. Similarly, the surface of the skin may be naked (as in moray eels), or covered with scales of a variety of different types usually defined as placoid (typical of sharks and rays), cosmoid (fossil lungfish and coelacanths), ganoid (various fossil fish but also living gars and bichirs), cycloid, and ctenoid (these last two are found on most bony fish). There are even fish that live mostly on land. Mudskippers feed and interact with one another on mudflats and go underwater to hide in their burrows. The catfish *Phreatobius cisternarum* lives in underground, phreatic habitats, and a relative lives in waterlogged leaf litter.

Fish range in size from the huge 16-metre (52 ft) whale shark to the tiny 8-millimetre (0.3 in) stout infantfish.

Fish species diversity is roughly divided equally between marine (oceanic) and freshwater ecosystems. Coral reefs in the Indo-

Pacific constitute the centre of diversity for marine fishes, whereas continental freshwater fishes are most diverse in large river basins of tropical rainforests, especially the Amazon, Congo, and Mekong basins. More than 5,600 fish species inhabit Neotropical freshwaters alone, such that Neotropical fishes represent about 10% of all vertebrate species on the Earth.

Anatomy

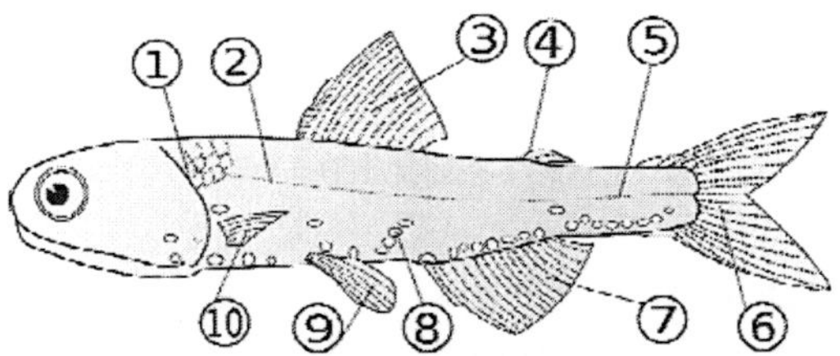

Figure: *The anatomy of* Lampanyctodes hectoris; *(1) operculum (gill cover), (2) lateral line, (3) dorsal fin, (4) fat fin, (5) caudal peduncle, (6) caudal fin, (7) anal fin, (8) photophores, (9) pelvic fins (paired), (10) pectoral fins (paired)*

Respiration

Most fish exchange gases using gills on either side of the pharynx. Gills consist of threadlike structures calledfilaments. Each filament contains a capillary network that provides a large surface area for exchanging oxygen andcarbon dioxide. Fish exchange gases by pulling oxygen-rich water through their mouths and pumping it over their gills. In some fish, capillary blood flows in the opposite direction to the water, causing countercurrent exchange. The gills push the oxygen-poor water out through openings in the sides of the pharynx. Some fish, like sharks and lampreys, possess multiple gill openings. However, bony fish have a single gill opening on each side. This opening is hidden beneath a protective bony cover called an operculum.

Juvenile bichirs have external gills, a very primitive feature that they share with larval amphibians.

Fish from multiple groups can live out of the water for extended time periods. Amphibious fish such as the mudskippercan live and move about on land for up to several days, or live in stagnant or otherwise oxygen depleted water. Many such fish can breathe air via a variety of mechanisms. The skin of anguillid eels may absorb oxygen directly. The buccal cavity of the electric eel may breathe air. Catfish of the families Loricariidae, Callichthyidae, and Scoloplacidae absorb air through their digestive tracts. Lungfish, with the exception of the

Australian lungfish, and bichirs have paired lungs similar to those of tetrapods and must surface to gulp fresh air through the mouth and pass spent air out through the gills. Gar andbowfin have a vascularised swim bladder that functions in the same way. Loaches, trahiras, and many catfish breathe by passing air through the gut. Mudskippers breathe by absorbing oxygen across the skin (similar to frogs). A number of fish have evolved so-called accessory breathing organs that extract oxygen from the air. Labyrinth fish (such as gouramis andbettas) have a labyrinth organ above the gills that performs this function. A few other fish have structures resembling labyrinth organs in form and function, most notably snakeheads,pikeheads, and the Clariidae catfish family.

Breathing air is primarily of use to fish that inhabit shallow, seasonally variable waters where the water's oxygen concentration may seasonally decline. Fish dependent solely on dissolved oxygen, such as perch and cichlids, quickly suffocate, while air-breathers survive for much longer, in some cases in water that is little more than wet mud. At the most extreme, some air-breathing fish are able to survive in damp burrows for weeks without water, entering a state of aestivation (summertime hibernation) until water returns.

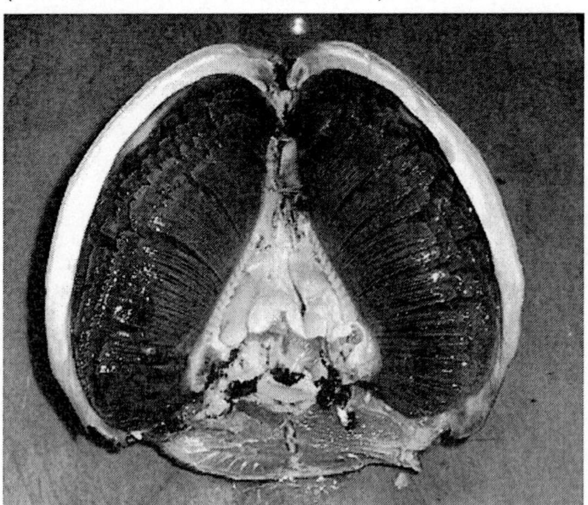

Figure: *Tuna gills inside of the head. The fish head is oriented snout-downwards, with the view looking towards the mouth.*

Air breathing fish can be divided into obligate air breathers and facultative air breathers. Obligate air breathers, such as theAfrican lungfish, *must* breathe air periodically or they suffocate. Facultative air breathers, such as the catfish *Hypostomus plecostomus*, only breathe air if they need to and will otherwise rely on their gills for oxygen.

Most air breathing fish are facultative air breathers that avoid the energetic cost of rising to the surface and the fitness cost of exposure to surface predators.

Circulation

Fish have a closed-loop circulatory system. The heart pumps the blood in a single loop throughout the body. In most fish, the heart consists of four parts, including two chambers and an entrance and exit. The first part is the sinus venosus, a thin-walled sac that collects blood from the fish's veins before allowing it to flow to the second part, the atrium, which is a large muscular chamber.

The atrium serves as a one-way antechamber, sends blood to the third part, ventricle. The ventricle is another thick-walled, muscular chamber and it pumps the blood, first to the fourth part, bulbus arteriosus, a large tube, and then out of the heart. The bulbus arteriosus connects to the aorta, through which blood flows to the gills for oxygenation.

Digestion

Jaws allow fish to eat a wide variety of food, including plants and other organisms. Fish ingest food through the mouth and break it down in the esophagus. In the stomach, food is further digested and, in many fish, processed in finger-shaped pouches calledpyloric caeca, which secrete digestive enzymes and absorb nutrients. Organs such as the liver and pancreas add enzymes and various chemicals as the food moves through the digestive tract. The intestine completes the process of digestion and nutrient absorption.

Excretion

As with many aquatic animals, most fish release their nitrogenous wastes as ammonia. Some of the wastes diffuse through the gills. Blood wastes are filtered by the kidneys. Saltwater fish tend to lose water because of osmosis. Their kidneys return water to the body. The reverse happens in freshwater fish: they tend to gain water osmotically. Their kidneys produce dilute urine for excretion. Some fish have specially adapted kidneys that vary in function, allowing them to move from freshwater to saltwater.

Scales

The scales of fish originate from the mesoderm (skin); they may be similar in structure to teeth.

Sensory and Nervous System

Central Nervous System: Fish typically have quite small brains relative to body size compared with other vertebrates, typically one-fifteenth the brain mass of a similarly sized bird or mammal. However, some fish have relatively large brains, most notably mormyrids and sharks, which have brains about as massive relative to body weight as birds and marsupials.

Fish brains are divided into several regions. At the front are the olfactory lobes, a pair of structures that receive and process signals from the nostrils via the two olfactory nerves. The olfactory lobes are very large in fish that hunt primarily by smell, such as hagfish, sharks, and catfish. Behind the olfactory lobes is the two-lobed telencephalon, the structural equivalent to the cerebrum in higher vertebrates. In fish the telencephalon is concerned mostly with olfaction. Together these structures form the forebrain.

Connecting the forebrain to the midbrain is the diencephalon (in the diagram, this structure is below the optic lobes and consequently not visible). The diencephalon performs functions associated with hormones and homeostasis. The pineal body lies just above the diencephalon. This structure detects light, maintains circadian rhythms, and controls colour changes. The midbrain or mesencephalon contains the two optic lobes. These are very large in species that hunt by sight, such as rainbow trout and cichlids.

The hindbrain or metencephalon is particularly involved in swimming and balance. The cerebellum is a single-lobed structure that is typically the biggest part of the brain. Hagfish and lampreys have relatively small cerebellae, while the mormyrid cerebellum is massive and apparently involved in their electrical sense. The brain stem or myelencephalon is the brain's posterior. As well as controlling some muscles and body organs, in bony fish at least, the brain stem governs respiration and osmoregulation.

Sense Organs

Most fish possess highly developed sense organs. Nearly all daylight fish have colour vision that is at least as good as a human's. Many fish also have chemoreceptors that are responsible for extraordinary senses of taste and smell. Although they have ears, many fish may not hear very well. Most fish have sensitive receptors that form the lateral line system, which detects gentle currents and vibrations, and senses the motion of nearby fish and prey. Some fish, such as catfish and sharks, have organs that detect weak electric currents on the order of millivolt. Other fish, like the South American electric fishes Gymnotiformes, can produce weak electric currents, which they use in navigation and social communication. Fish orient themselves using landmarks and may use mental maps based on multiple landmarks or symbols. Fish behaviour in mazes reveals that they possess spatial memory and visual discrimination.

Capacity for Pain

Experiments done by William Tavolga provide evidence that fish have pain and fear responses. For instance, in Tavolga's experiments, toadfish grunted when electrically shocked and over time they came to grunt at the mere sight of an electrode.

In 2003, Scottish scientists at the University of Edinburgh and the Roslin Institute concluded that rainbow trout exhibit behaviours often associated with pain in other animals. Bee venomand acetic acid injected into the lips resulted in fish rocking their bodies and rubbing their lips along the sides and floors of their tanks, which the researchers concluded were attempts to relieve pain, similar to what mammals would do. Neurons fired in a pattern resembling human neuronal patterns.

Professor James D. Rose of the University of Wyoming claimed the study was flawed since it did not provide proof that fish possess "conscious awareness, particularly a kind of awareness that is meaningfully like ours". Rose argues that since fish brains are so different from human brains, fish are probably not conscious in the manner humans are, so that reactions similar to human reactions to pain instead have other causes. Rose had published a study a year earlier arguing that fish cannot feel pain because their brains lack aneocortex. However, animal behaviourist Temple Grandin argues that fish could still have consciousness without a neocortex because "different species can use different brain structures and systems to handle the same functions."

Animal welfare advocates raise concerns about the possible suffering of fish caused by angling. Some countries, such as Germany have banned specific types of fishing, and the British RSPCA now formally prosecutes individuals who are cruel to fish.

Muscular System

Most fish move by alternately contracting paired sets of muscles on either side of the backbone. These contractions form S-shaped curves that move down the body. As each curve reaches the back fin, backward force is applied to the water, and in conjunction with the fins, moves the fish forward. The fish's fins function like an airplane's flaps. Fins also increase the tail's surface area, increasing speed. The streamlined body of the fish decreases the amount of friction from the water. Since body tissue is denser than water, fish must compensate for the difference or they will sink. Many bony fish have an internal organ called a swim bladder that adjusts their buoyancy through manipulation of gases.

Figure: *A 3-tonne (3.0-long-ton; 3.3-short-ton)great white shark off Isla Guadalupe*

Homeothermy

Although most fish are exclusively ectothermic, there are exceptions. Certain species of fish maintain elevated body temperatures. Endothermic teleosts (bony fish) are all in the suborder Scombroidei and include the billfishes, tunas, and one species of "primitive" mackerel (*Gasterochisma melampus*). All sharks in the family Lamnidae – shortfin mako, long fin mako, white, porbeagle, and salmon shark – are endothermic, and evidence suggests the trait exists in family Alopiidae (thresher sharks). The degree of endothermy varies from the billfish, which warm only their eyes and brain, to bluefin tuna and porbeagle sharks who maintain body temperatures elevated in excess of 20 °C above ambient water temperatures. Endothermy, though metabolically costly, is thought to provide advantages such as increased muscle strength, higher rates of central nervous system processing, and higher rates of digestion.

Reproductive System

Organs

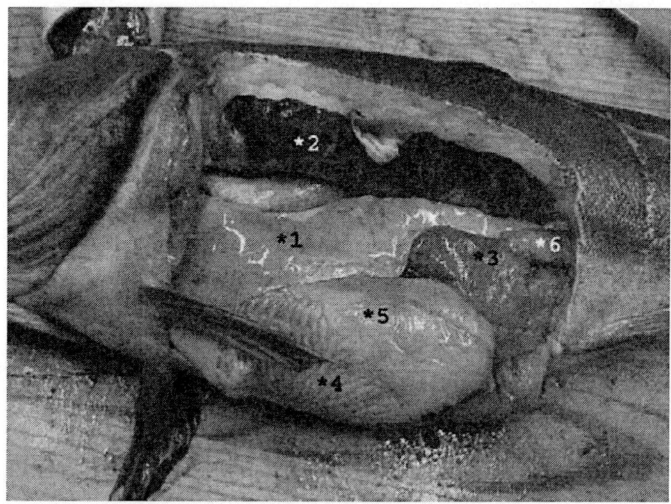

Figure: *Organs: 1. Liver, 2. Gas bladder, 3. Roe, 4. Pyloric caeca, 5. Stomach, 6. Intestine*

Fish reproductive organs include testes and ovaries. In most species, gonads are paired organs of similar size, which can be partially or totally fused. There may also be a range of secondary organs that increase reproductive fitness. In terms of spermatogonia distribution, the structure of teleosts testes has two types: in the most common, spermatogonia occur all along theseminiferous tubules, while in Atherinomorph fish they are confined to the distal portion of these structures. Fish can present cystic or semi-cystic spermatogenesis in relation to the release phase of germ cells in cysts to the seminiferous tubules lumen.

Fish ovaries may be of three types: gymnovarian, secondary gymnovarian or cystovarian. In the first type, the oocytes are released directly into the coelomic cavity and then enter the ostium, then through the oviduct and are eliminated. Secondary gymnovarian ovaries shed ova into thecoelom from which they go directly into the oviduct. In the third type, the oocytes are conveyed to the exterior through the oviduct.Gymnovaries are the primitive condition found in lungfish, sturgeon, and bowfin. Cystovaries characterise most teleosts, where the ovary lumen has continuity with the oviduct. Secondary gymnovaries are found in salmonids and a few other teleosts. Oogonia development in teleosts fish varies according to the group, and the determination of oogenesis dynamics allows the understanding of maturation and fertilization processes. Changes in the nucleus, ooplasm, and the surrounding layers characterise the oocyte maturation process.

Postovulatory follicles are structures formed after oocyte release; they do not have endocrine function, present a wide irregular lumen, and are rapidly reabsorbed in a process involving the apoptosis of follicular cells. A degenerative process called follicular atresia reabsorbs vitellogenic oocytes not spawned. This process can also occur, but less frequently, in oocytes in other development stages. Some fish are hermaphrodites, having both testes and ovaries either at different phases in their life cycle or, as in hamlets, have them simultaneously.

Reproductive Method

Over 97% of all known fish are oviparous, that is, the eggs develop outside the mother's body. Examples of oviparous fish include salmon, goldfish, cichlids, tuna, and eels. In the majority of these species, fertilisation takes place outside the mother's body, with the male and female fish shedding their gametes into the surrounding water. However, a few oviparous fish practice internal fertilization, with the male using some sort of intromittent organ to deliver sperm into the genital opening of the female, most notably the oviparous sharks, such as the horn shark, and oviparous rays, such as skates. In these cases, the male is equipped with a pair of modified pelvic fins known as claspers. Marine fish can produce high numbers of eggs which are often released into the open water column. The eggs have an average diametre of 1 millimetre (0.039 in).

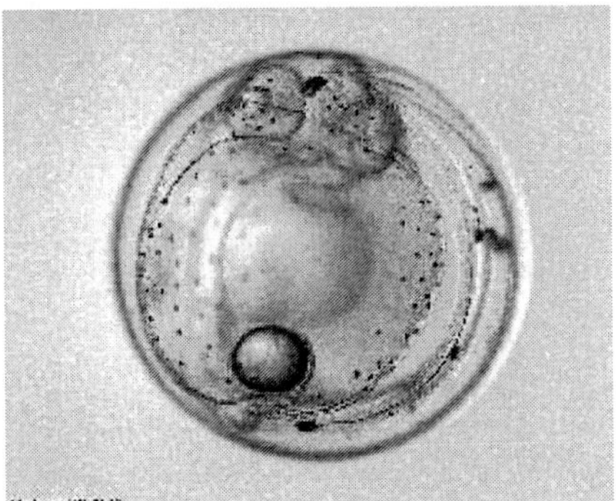

Figure: *An example of zooplankton*

The newly hatched young of oviparous fish are called larvae. They are usually poorly formed, carry a large yolk sac (for nourishment) and are very different in appearance from juvenile and adult specimens. The larval period in oviparous fish is relatively short (usually only several weeks), and larvae rapidly grow and change appearance and structure (a process termed metamorphosis) to become juveniles. During this transition larvae must switch from their yolk sac to feeding on zooplankton prey, a process which depends on typically inadequate zooplankton density, starving many larvae.

In ovoviviparous fish the eggs develop inside the mother's body after internal fertilization but receive little or no nourishment directly from the mother, depending instead on the yolk. Each embryo develops in its own egg. Familiar examples of ovoviviparous fish include guppies, angel sharks, and coelacanths.

Some species of fish are viviparous. In such species the mother retains the eggs and nourishes the embryos. Typically, viviparous fish have a structure analogous to the placenta seen in mammals connecting the mother's blood supply with that of the embryo. Examples of viviparous fish include the surf-perches, splitfins, and lemon shark. Some viviparous fish exhibit oophagy, in which the developing embryos eat other eggs produced by the mother. This has been observed primarily among sharks, such as the shortfin mako and porbeagle, but is known for a few bony fish as well, such as the halfbeak *Nomorhamphus ebrardtii*. Intrauterine cannibalism is an even more unusual mode of vivipary, in which the largest embryos eat weaker and smaller siblings. This behaviour is also most commonly found among sharks, such as the grey nurse shark, but has also been reported for *Nomorhamphus ebrardtii*. Aquarists commonly refer to ovoviviparous and viviparous fish as livebearers.

Immune System

Immune organs vary by type of fish. In the jawless fish (lampreys and hagfish), true lymphoid organs are absent. These fish rely on regions of lymphoid tissue within other organs to produce immune cells. For example, erythrocytes, macrophages and plasma cells are produced in the anterior kidney (or pronephros) and some areas of the gut (where granulocytesmature.) They resemble primitive bone marrow in hagfish. Cartilaginous fish (sharks and rays) have a more advanced immune system. They have three specialised organs that are unique to chondrichthyes; the epigonal organs (lymphoid tissue similar to mammalian bone) that surround the gonads, the Leydig's organ within the walls of their esophagus, and a spiral valve in their intestine. These organs house typical immune cells (granulocytes, lymphocytes and plasma cells). They also possess an identifiable thymus and a well-developed spleen (their most important immune organ) where various lymphocytes, plasma cells and macrophages develop and are stored. Chondrostean fish (sturgeons, paddlefish and bichirs) possess a major site for the production of granulocytes within a mass that is associated with the meninges (membranes surrounding the central nervous system.) Their heart is frequently covered with tissue that contains lymphocytes, reticular cells and a small number of macrophages. The chondrostean kidney is an important hemopoietic organ; where erythrocytes, granulocytes, lymphocytes and macrophages develop.

Like chondrostean fish, the major immune tissues of bony fish (or teleostei) include the kidney (especially the anterior kidney), which houses many different immune cells. In addition, teleost fish possess a thymus, spleen and scattered immune areas within mucosal tissues (e.g. in the skin, gills, gut and gonads). Much like the mammalian immune system, teleost erythrocytes, neutrophils and granulocytes are believed to reside in the spleen whereas lymphocytes are the major cell type found in the thymus. In 2006, a lymphatic system similar to that in mammals was described in one species of teleost fish, the zebrafish. Although not confirmed as yet, this system presumably will be where naive (unstimulated) T cellsaccumulate while waiting to encounter an antigen.

Diseases

Like other animals, fish suffer from diseases and parasites. To prevent disease they have a variety of defences. *Non-specific* defences include the skin and scales, as well as the mucus layer secreted by the epidermis that traps and inhibits the growth of microorganisms. If pathogens breach these defences, fish can develop an inflammatory response that increases blood flow to the infected region and delivers white blood cells that attempt to destroy pathogens. Specific defences respond to particular pathogens recognised by the fish's body, i.e., animmune response. In recent years, vaccines have become widely used in aquaculture and also with ornamental fish, for example furunculosis vaccines in farmed salmon and koi herpes virus in koi.

Some species use cleaner fish to remove external parasites. The best known of these are the Bluestreak cleaner wrasses of the genus *Labroides* found on coral reefs in the Indian andPacific Oceans. These small fish maintain so-called "cleaning stations" where other fish congregate and perform specific movements to attract the attention of the cleaners. Cleaning behaviours have been observed in a number of fish groups, including an interesting case between two cichlids of the same genus, *Etroplus maculatus*, the cleaner, and the much larger*Etroplus suratensis*.

Conservation

The 2006 IUCN Red List names 1,173 fish species that are threatened with extinction. Included are species such as Atlantic cod, Devil's Hole pupfish, coelacanths, and great white sharks. Because fish live underwater they are more difficult to study than terrestrial animals and plants, and information about fish populations is often lacking. However, freshwater fish seem particularly threatened because they often live in relatively small water bodies. For example, the Devil's Hole pupfish occupies only a single 3 by 6 metres (10 by 20 ft) pool.

Overfishing

Figure: *A Whale shark, the world's largest fish, is classified as Vulnerable.*

Overfishing is a major threat to edible fish such as cod and tuna. Overfishing eventually causes population (known as stock) collapse because the survivors cannot produce enough young to replace those removed. Such commercial extinction does not mean that the species is extinct, merely that it can no longer sustain a fishery.

One well-studied example of fishery collapse is the Pacific sardine *Sadinops sagax caerulues* fishery off the California coast. From a 1937 peak of 790,000 long tons (800,000 t) the catch steadily declined to only 24,000 long tons (24,000 t) in 1968, after which the fishery was no longer economically viable.

The main tension between fisheries science and the fishing industry is that the two groups have different views on the resiliency of fisheries to intensive fishing. In places such as Scotland, Newfoundland, and Alaska the fishing industry is a major employer, so governments are predisposed to support it. On the other hand, scientists and conservationists push for stringent protection, warning that many stocks could be wiped out within fifty years.

Importance to Humans

Culture

In the Book of Jonah a "great fish" swallowed Jonah the Prophet. Legends of half-human, half-fish mermaids have featured in stories like those of Hans Christian Andersen and movies like *Splash*. Among the deities said to take the form of a fish are Ika-Roa of the Polynesians, Dagon of various ancient Semitic peoples, the shark-gods of Hawaii and Matsya of the Hindus. The astrological symbol Pisces is based on a constellation of the same name, but there is also a second fish constellation in the night sky, Piscis Austrinus. Fish have been used figuratively in many different ways, for example the ichthys used by early Christians to identify themselves, through to the fish as a symbol of fertility among Bengalis. Fish feature prominently in art and literature, in movies such as *Finding Nemo* and books such as *The Old Man and the Sea*. Large fish, particularly sharks, have frequently been the subject of horror movies and thrillers, most notably the novel *Jaws*, which spawned a series of films of the same namethat in turn inspired similar films or parodies such as *Shark Tale*, *Snakehead Terror*, and *Piranha*.

In the semiotic of Ashtamangala (buddhist symbolism) the golden fish (Sanskrit: Matsya), represents the state of fearless suspension in samsara, perceived as the harmless ocean, referred to as 'buddha-eyes' or 'rigpa-sight'. The fish symbolises the auspiciousness of all living beings in a state of fearlessness without danger of drowning in the Samsaric Ocean of Suffering, and migrating from teaching to teaching freely and spontaneously just as fish swim. They have religious significance in Hindu, Jain and Buddhist traditions but also in Christianity who is first signified by the sign of the fish, and especially referring to feeding the multitude in the desert. In the dhamma of Buddha the fish symbolise happiness as they have complete freedom of movement in the water. They represent fertility and abundance. Often drawn in the form of carp which are regarded in the Orient as sacred on account of their elegant beauty, size and life-span. The name of the Canadian city of Coquitlam, British Columbia is derived from *Kwikwetlem*, which is said to be derived from a Coast Salish term meaning "little red fish".

Terminology

Shoal or School?

A random assemblage of fish merely using some localised resource such as food or nesting sites is known simply as an *aggregation*. When fish come together in an interactive, social grouping, then they may be forming either a *shoal* or a *school* depending on the degree of organisation. A *shoal* is a loosely organised group where each fish swims and forages independently but is attracted to other members of the group and adjusts its behaviour, such as swimming speed, so that it remains close to the other members of the group. *Schools* of fish are much more tightly organised, synchronising their swimming so that all fish move at the same speed and in the same direction. Shoaling and schooling behaviour is believed to provide a variety of advantages.

Examples:

- Cichlids congregating at lekking sites form an *aggregation.*
- Many minnows and characins form *shoals.*
- Anchovies, herrings and silversides are classic examples of *schooling* fish.

While school and shoal have different meanings within biology, they are often treated as synonyms by non-specialists, with speakers of British English using "shoal" to describe any grouping of fish, while speakers of American English often using "school" just as loosely.

Fish or Fishes?

Though often used interchangeably, these words have different meanings. *Fish* is used either as singular noun or to describe a group of specimens from a single species. *Fishes*describes a group of different species:

> *"The present encyclopaedic work on fishes is designed to impart latest information in the light of recent developments and researches in the field besides the basic conventional information.*

This text is designed to approach the morphology, anatomy, physiology and development of the Fishes in a coherent way since it is most natural for a student in studying the architectural elements of an animal body in the quest to know how they function.

Profusely illustrated, this work features a text, reference and a laboratory guide. It deals with the structures and physiological phenomena of each system. Origin, Adaptative radiation and other general topics are also dealt in details. No doubt, this work, in three volumes, will not only meet the requirement of Indian students but will also be useful as a guid.

Adaptations in Fishes

Missouri fishes inhabit many different aquatic habitats from riffles and pools of large rivers and small streams to natural and artificial lakes, ponds and oxbows. Each species has adapted to survive in one or more of these habitats, but few can survive in all habitats.

Traits that allow a species to live in a particular habitat can be adaptations in temperature tolerance, salt tolerance, the ability to breath air when needed, and in body form.

Adaptations of a fish's body that allow a species to survive in different habitats include shape of the body, fin location and mouth size and orientation. In general, body forms of Missouri fishes can be grouped into seven different categories: rover predator, lie-in-wait predator, deep-bodied fish, eel-like fish, bottom rover, bottom clinger and surface-oriented fish.

Rover Predator

Temperate and black basses (like the largemouth bass illustrated above), trout and walleye are examples of fish with body shapes adapted to a rover-predator lifestyle. Their bodies are elongate, their fins are distributed around the body evenly, their head is pointed with a relatively large mouth at the tip, and they usually have a forked tail, an indicator of a fast swimmer. This type of fish roves through several types of habitats from still to swift waters, and uses speed to catch prey. Many of our favourite Missouri game fish are rover predators.

Lie-in-wait predator

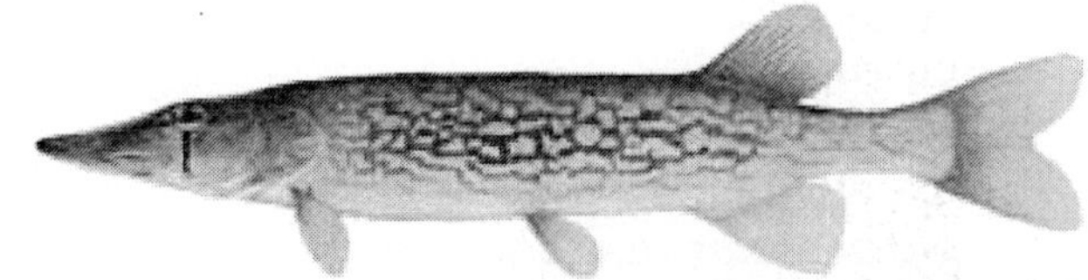

Pickerel (illustrated above) and gar are Missouri species with the lie-in-wait predator body adaptation. Their bodies are elongate, their fins are moved back on the body to provide thrust for chasing prey, their head is pointed and flattened on top, and they have a mouth full of teeth at the tip. This type of fish will lie hidden by cover in still or slow moving waters and dart out to snap up prey.

Deep-bodied Fish

Many sunfish (like the bluegill illustrated above) and buffalofish species exhibit the deep-bodied form. Deep-bodied fishes are flattened laterally and have long dorsal and anal fins.

Their pectoral and pelvic fins are located high on the body. This type of fish mainly occurs in still waters and are adapted for maneuverability in heavy cover. They typically feed from the bottom or pick slow-moving prey from the water column.

Eel-like Fish

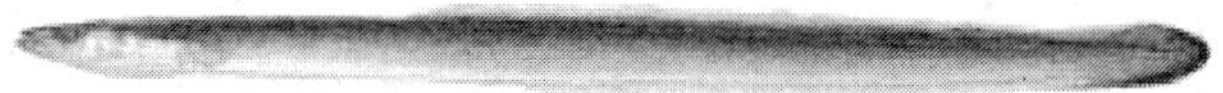

American eel (illustrated above) and lampreys are examples of Missouri eel-like fish. They have very elongate bodies with reduced pelvic and pectoral fins, very long anal and dorsal fins, and unforked tails. This type of fish occurs in still to moderately swift waters and is adapted for maneuverability in tight places and burrowing in soft substrates. Eels typically sneak up on prey, and lampreys filter food from the bottom with some species becoming parasitic as adults.

Bottom Clinger

Sculpins (like the banded sculpin illustrated above) and many darter species are adapted as bottom clingers. They tend to have flattened heads and large pectoral fins that are angled to keep the fish on the bottom in swift currents. This type of fish occurs in the swift water of riffles where they pick invertebrates from the rocks

Bottom Rover

Many Missouri catfish and suckers (like the golden redhorse illustrated above) are adapted as bottom rovers. These species have a body shape similar to the rover predator except the back is humped, the head is flattened, and the pectoral fins are enlarged. This type of fish occurs in still to swift waters and uses speed to avoid predators. The mouths of bottom rovers are placed in several different positions depending on their feeding style. Omnivours like catfish have a terminal mouth that allows them to take prey if the opportunity arises, generalist bottom feeders like many suckers have a subterminal mouth, and bottom feeders like carp have ventral mouths with protrusible lips for sucking ooze. Many bottom rovers also have well developed barbels for locating food by feel.

Surface-oriented

Missouri's topminnows (like the blackstripe topminnow illustrated above) and studfish are surface-oriented species. They are typically small-bodied and have a flat head with upturned mouth. Their dorsal fin is moved back on their body. This type of fish occurs in still waters and feeds on organisms that fall in the water. Many surface-oriented species can survive in deoxygenated waters by breathing the surface layer of water that always contains some oxygen.

How Fish Use Adaptations

Colouration

Fish display a wide variety of colours and colour patterns. Skin colouration can have many functions. Many fish have colour patterns that help them blend in with their environment. This may allow the fish to avoid being seen by a predator. Some fish, such as the flat fishes (Pleuronectiformes), can change their skin colouration to match the surrounding habitat.

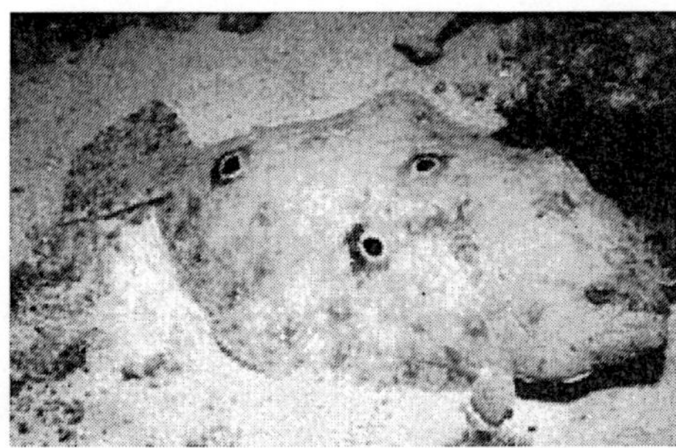

Figure: *Gulf flounder*

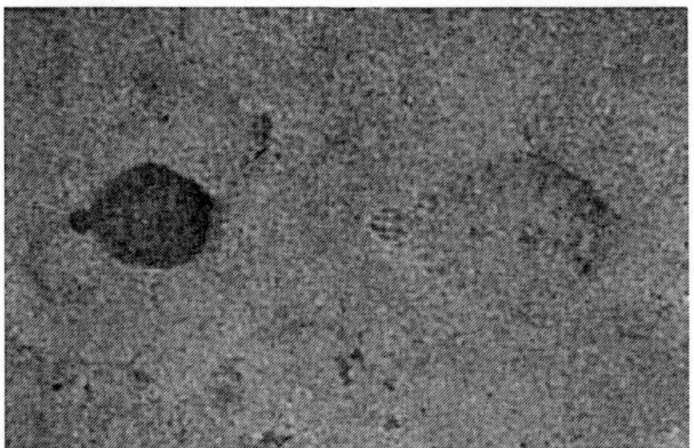

Figure: *Wide-eyed flounder.*

Fish can also have disruptive markings to hide body parts. Species such as the jackknife fish (*Equetus lanceolatus*), high-hat (*Equetus acuminatus*) and some angel fishes (Pomacanthidae), have dark lines that run through the eyes. These lines may serve to hide the eyes so that other animals can not tell where the fish is looking or even if it is a fish. Also, horizontal lines may be a sight-line for aiming attacks on prey. Some fishes, like butterflyfishes (Chaetodontidae), have spots on their body that resemble eyes. This may serve to confuse prey and predators alike. In addition to colouration, some fish, like the sea dragon (*Phyllopteryx*), have body shapes that can further mimic their habitat.

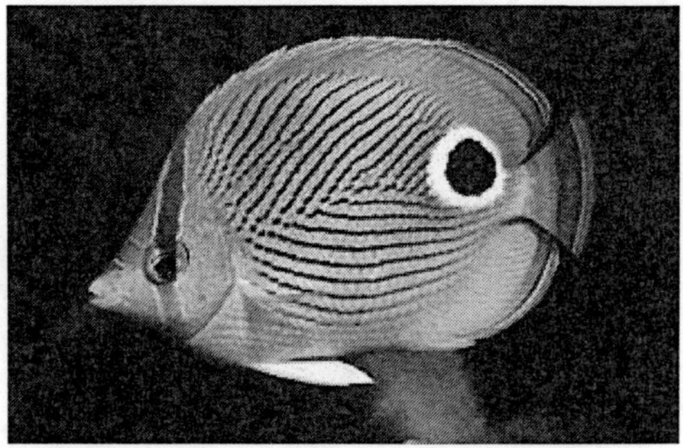

Figure: *Foureye butterflyfish.*

Figure: *High-hat, family Sciaenidae* David Snyder

Fish colouration can also be useful in catching prey. Many sharks exhibit colouration known as counter shading. Sharks that have counter shading are dark on the dorsal (upper) side and light on the ventral (lower) side. With this colour scheme any prey looking down on the shark will see a dark shark against a dark sea bottom, making it hard to detect the shark. Conversely, any prey looking up at the shark, will see the light belly of the shark on the light background of the ocean surface water lit by the sun or moon.

Figure: *Gray reef shark.*

Figure: *Silvertip shark.*

Colouration can also be used to advertise. Fishes like the darters (Percidae) and sticklebacks (*Gasterosteus*), may use colour to attract and recognise potential mates.

Figure: *Rainbow darter.*

Figure: *Coosa darter.*

Light Organs

Some marine fish have the ability to produce light through bioluminescence. Most light-producing fish live in mid-water or are bottom-dwelling deep sea species.

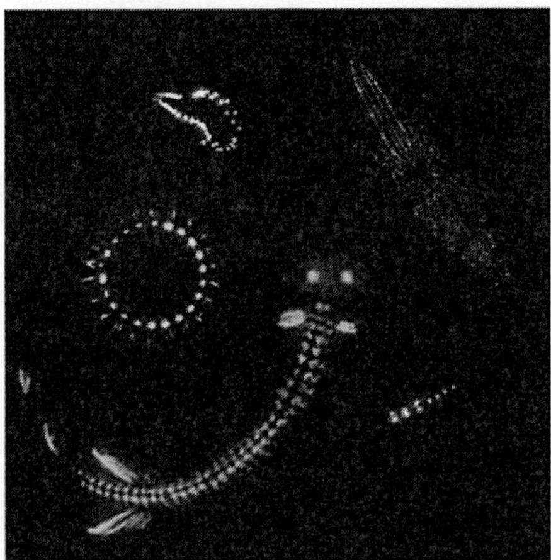

Figure: *Deep water bioluminescent organisms* courtesy NOAA

In fish, bioluminescence can occur two different ways: through symbiotic bacteria living on the fish or through self-luminous cells called photophores. Some species of deep sea angler fish (Lophiiformes) may use this light to attract prey, while others, like the Atlantic midshipman (*Porichthys plectrodon*), may use this light to attract mates.

Venom

Many fish may use venom as a form of defence. Most venomous fish deliver the toxins through the use of a spine. Venomous spines are found in a wide variety of fish including stingrays, chimaeras, scorpionfishes, catfishes, toadfishes, rabbit fishes, and stargazers. Venomous spines can have poison glands along the grove of the spine, as with stingrays, or at the base of the spine, as in some catfish. While humans can be stung by a multitude of fishes, few species are life-threatening.

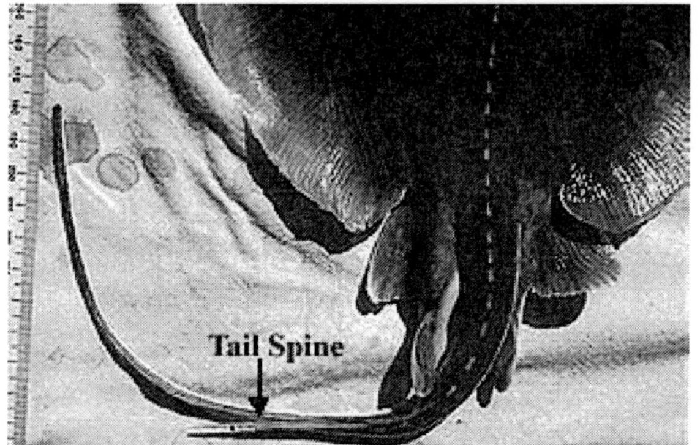

Figure: *Stingray tail spine.*

Electric Organs

Elasmobranchs (sharks, skates, and rays) possess an electric sense system known as the ampullae of Lorenzini. This system consists of many tiny gel-filled canals positioned on the head of the fish. Through this system these fishes are able to detect the weak electric fields produced by prey. It is also believed that these fish can use this sense to detect the electric fields they induce when swimming through the earth's magnetic field, as a sort of compass. Since the fishes are able to generate the fields they detect, this is a form of active electro-orientation. Some species of skates and rays also have electricity-producing organs. The electric rays have paired electric organs located on either side of the head, behind the eyes. With these organs, electric rays are able to shock and stun their prey. The skate's electric organs are located near the tail. However, these electric organs only produce weak electric fields not capable of stunning prey. Researchers believe that the skate's electric organs are used for communication and mate location.

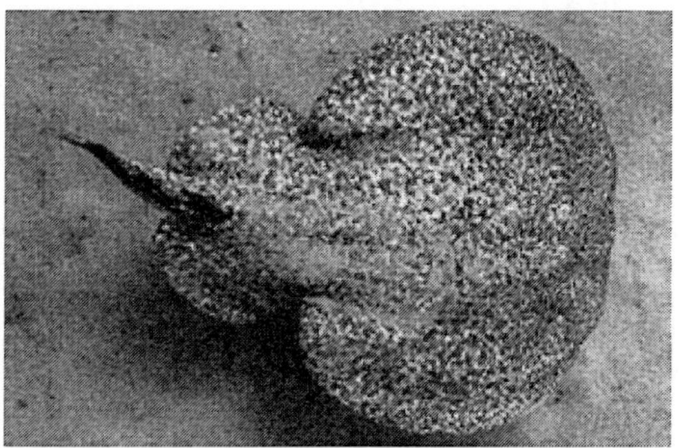

***Figure:** Marbled electric ray.*

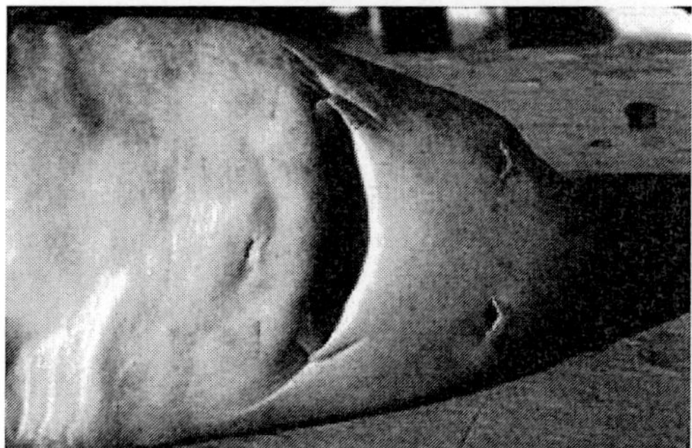

***Figure:** Roughskin spiny dogfish.*

The electric eel can also produce electric fields. These eels use weak electric fields for navigation, prey location, and communication. Additionally, these eels can produce strong electric fields to stun potential prey. The strength of the "shock" is related to the size of the eel, with larger individuals being able to produce more of a "shock."

***Figure:** Electric eel.*

Origin and Evolution of Fishes

The History of Animal Evolution

For many people animals are perhaps the most familiar, and most interesting, of living things. This may be because we are animals ourselves. As such, we have a number of features in common with all the organisms placed in the animal kingdom, and these common features indicate that we have a shared evolutionary history.

All animals and plants are classified as multicellular eukaryotes: their bodies are made up of large numbers of cells, and microscopic inspection of these cells reveals that they contain a nucleus and a number of other organelles. Compared to prokaryotic organisms such as bacteria, plants and animals have a relatively recent evolutionary origin. DNA evidence suggests that the first eukaryotes evolved from prokaryotes, between 2500 and 1000 million years ago.

That is, eukaryotes as a taxon date from the Proterozoic Era, the final Era of the Precambrian. Fossils of both simple unicellular and more complex multicellular organisms are found in abundance in rocks from this period of time. In fact, the name "Proterozoic" means "early life".

Plants and animals both owe their origins to endosymbiosis, a process where one cell ingests another, but for some reason then fails to digest it. The evidence for this lies in the way their cells function. Both plant and animal rely on structures called mitochondria to release energy in their cells, using aerobic respiration to produce the energy-carrying molecule ATP.

There is considerable evidence that mitochondria evolved from free-living aerobic bacteria: they are the size of bacterial cells; they divide independently of the cell by binary fission; they have their own genome in the form of a single circular DNA molecule; their ribosomes are more similar to those of bacteria than to the ribosomes found in the eukaryote cell's cytoplasm; and like chloroplasts they are enclosed by a double membrane as would be expected if they derived from bacterial cells engulfed by another cell.

Like the plants, animals evolved in the sea. And that is where they remained for at least 600 million years. This is because, in the absence of a protective ozone layer, the land was bathed in lethal levels of UV radiation. Once photosynthesis had raised atmospheric oxygen levels high enough, the ozone layer formed, meaning that it was then possible for living things to venture onto the land.

The oldest fossil evidence of multicellular animals, or metazoans, is burrows that appear to have been made by smooth, wormlike organisms. Such trace fossils have been found in rocks from China, Canada, and India, but they tell us little about the animals that made them apart from their basic shape.

- The Ediacaran animals
- The Cambrian "explosion" and the Burgess Shale
- What caused the Cambrian "explosion"?
- A foot on the land
- The earliest vertebrates
- Appearance of the fish
- The jawless fish.

A Foot on the Land

Whatever their origins, animals may have ventured onto land early in the Cambrian. Previously scientists believed that animals did not begin to colonise the land until the Silurian (440 - 410 million years ago). However, the 2002 discovery of the footprints of animals that scuttled about on sand dunes about 530 million years ago has changed this view. These animals were arthropods, and resembled centipedes about the size of crayfish. They probably didn't live on land, instead coming ashore to mate or evade predators. At this time the only land plants appear to have resembled mosses .

The Jawless Fish

Agnathans, or jawless fish, were the earliest fish: an excellent fossil of *Haikouichthys ercaicunensis* dates back about 530 million years, to the Cambrian. Previously the earliest-known agnathans were dated to around 480 million years ago. Agnathans have traditionally been placed with the vertebrates due to the presence of a skull, although the modern forms such as hagfish lack a vertebral column. The earliest agnathans were Ostracoderms. They were bottom-feeders and were almost entirely covered in armour plates. When the sharks and bony fish began to evolve, around 450 million years ago, most ostracoderms became extinct. Only the lineage that produced the modern hagfish and lampreys survived.

Fish Evolution

The first fishes, and indeed the first vertebrates, were the ostracoderms, which appeared in the Cambrian Period, about 510 million years ago, and became extinct at the end of the Devonian, about 350 million years ago. Ostracoderms were jawless fishes found mainly in fresh water. They were covered with a bony armour or scales and were often less than 30 cm (1 ft) long. The ostracoderms are placed in the class Agnatha along with the living jawless fishes, the lampreys and hagfishes, which are believed to be descended from the ostracoderms.

The first fishes with jaws, the acanthodians, or spiny sharks, appeared in the late Silurian, about 410 million years ago, and became extinct before the end of the Permian, about 250 million years ago. Acanthodians were generally small sharklike fishes varying from toothless filter-feeders to toothed predators. They are often classified as an order of the class Placodermi, another group of primitive fishes, but recent authorities tend to place the acanthodians in a class by themselves (class Acanthodii) or even within the class of modern bony fishes, the Osteichthyes. It is commonly believed that the acanthodians and the modern bony fishes are related and that either the acanthodians gave rise to the modern bony fishes or that both groups share a common ancestor. The placoderms, another group of jawed fishes, appeared at the beginning of the Devonian, about 395 million years ago, and became extinct at the end of the Devonian or the beginning of the Mississippian (Carboniferous), about 345 million years ago. Placoderms were typically small, flattened bottom-dwellers. The upper jaw was firmly fused to the skull, but there was a hinge joint between the skull and the bony plating of the trunk region.

The cartilaginous-skeleton sharks and rays, class Chondrichthyes, which appeared about 370 million years ago in the middle Devonian, are generally believed to be descended from the bony-skeleton placoderms. The cartilaginous skeletons are considered to be a later development. The modern bony fishes, class Osteichthyes, appeared in the late Silurian or early Devonian, about 395 million years ago. The early forms were freshwater fishes, for no fossil remains of modern bony fishes have been found in marine deposits older than Triassic time, about 230 million years ago. The Osteichthyes may have arisen from the acanthodians. A subclass of the Osteichthyes, the ray-finned fishes (subclass Actinopterygii), became and have remained the dominant group of fishes throughout the world. It was not the ray-finned fishes, however, that led to the evolution of the land vertebrates.

The ancestors of the land vertebrates are found among another group of bony fishes called the Choanichthyes or Sarcopterygii.

Choanate fishes are characterised by internal nostrils, fleshy fins called lobe fins, and cosmoid scales. The choanate fishes appeared in the late Silurian or early Devonian, more than 390 million years ago, and possibly arose from the acanthodians. The choanate fishes include a group known as the Crossopterygii, which has one living representative, the coelacanth Latimeria. During the Devonian Period some crossopterygian fishes of the order (or suborder) Rhipidistia crawled out of the water to become the first amphibians.

The Evolutionary Steps of Fish

Fish are incredible diverse group, made up of three living classes that hold an important place both in modern ecology and in evolutionary history. Fish are incredibly successful, they are numerous and have a huge variety of adaptations. They are successful in every water filled habitat. Furthermore, fish are the first known vertebrates and also the stepping stone to all land-walking vertebrates (tetrapods). Fish are complicated, diverse, and have a long evolutionary history.

In order to understand the place of fish as both a transitory group and as a modern one it is important to understand the times in which and from which they evolved. Fish are the first known true chordates. The first vertebrate that has been found is the Upper Cambrian fossil Anaspis, which is more than 500 million years old. This fossil, while being fragmentary, is thought to be an armoured, jawless fish. Fish did most of their evolving between five million and three and a half million years ago. These two periods were known as the Silurian and the Devonian periods. In yhe middle Silurian, the jawless fishes had diversified, but it was not until the Devonian that the true variety of fishes really flourished. In fact, the Devonian, is often referred to as the "Age of Fishes". Towards the end of the Devonian the first tetrapods (vertebrates which evolved true legs with which they could walk on land) had evolved from one specific branch of fish. Fish greatly specialised in their aquatic niche during both the Devonian and the Silurian and part of this evolution led to adaptations to land in the form of amphibians.

Fish are in no way simply a stepping stone to amphibians, they are a much more significant than this. They have evolved to be masters of their domain, the water. They come in many forms, have the ability to eat a huge variety of foods, and have populated almost every body of water. In fact, fish are the most common vertebrate, with there being approximately 24,000 species alive today. This number is mind boggling when put in perspective; the next most common vertebrate are birds with a mere 8,600 species. This multitude of species ranges in size, morphology, agility, and adaptations to environment. Fish have been broken down into a series of classes that separate them based on characteristics. These groups help to classify the wide variety of species that make up fish and help lead to understand of the current and evolutionary niche of fish. The first fish to evolve were the Agnathans (Class agnatha). These jawless fishes are the first vertebrates. These fish have round mouth parts that could be used for sucking or filter feeding. These rasping, sucking mouths are currently found on modern lampreys and hagfishes. These fish were often extremely armoured in order to help them protect themselves.

One group that evolved before the Silurian were the Ostracoderms which have been described as "small, blunt-headed forms they fed on debris in the mud, bullet shaped swimmers, and some with an astonishing array of spines and crests on their heads." Most of these types of fish are currently extinct with the exceptions of the lampreys and the hagfish. From these bottom feeding, jawless fish came the evolution of jawed fish. Jaws evolved only once (rather than evolving multiple times in different species through parallel evolution). Jaws evolved from gill arches which are the bony parts between gill slits. It is thought that a gill arch in an agnathan became fused to its skull. The upper part of the gill support became the top jaw and the bottom part of the gill support became the bottom jaw. Embryology points to this and the arrangement of nerves in shark heads and most simple fishes shows that jaws are in line with gill arches. While fish had the first bony jaws, they also have some of the most complicated. While the human head has only one moving part (the jaw), the head of a fish may have more than twenty-four bones that may move together in feeding. The evolution of the jaw is incredibly important because it led to fish to be able to ingest a much wider variety of foods and allowed them to be active hunters as opposed to passive filter feeder. This led to a wide variety of adaptations in morphology. Fish became more agile to be better predators, they were able to reduce their armour because they were less vulnerable, and their muscle density was able to decrease becuase they no longer led such a sluggish lifestyle.

The Evolutionary Origins of Syngnathidae: Pipefishes and Seahorses

The Syngnathidae (seahorses and pipefishes) are a large family of close to 300 marine, brackish and freshwater species (Froese & Pauly, 2010), all of which share an exceptional form of reproduction, male pregnancy. This key feature of the group has made them important model organisms for the study of sexual selection. In addition to their unique reproductive behaviour, syngnathids are well known for their highly specialised morphology, and the diversity of morphological forms found in this group has complicated efforts to understand their evolutionary origins and pattern of diversification. This exceptional morphological variation is reflected in the current taxonomy of the group: 14 of the 54 currently recognised syngnathid genera are monotypic (Froese & Pauly, 2010), and the majority of genera are composed of fewer than three species.

The family Syngnathidae has traditionally been included as a member of the order Gasterosteiformes, which includes 11 families in two suborders: the Gasterosteoidei, with the Hypoptychidae, Gasterosteidae and Aulorhynchidae; and the Syngnathoidei, with the Indostomidae, Aulostomidae, Fistulariidae, Macroramphosidae, Centriscidae, Pegasidae, Solenostomidae and Syngnathidae. The evolutionary history of the Gasterosteiformes itself has been the subject of controversy since its inception because of the derived and highly reductive morphology of its constituent families. Major monophyletic groups of families within the Gasterosteiformes have long been recognised, including Gasterosteidae and Aulorhynchidae (as superfamily Gasterosteoidea), Aulostomidae and Fistulariidae (Aulostomoidea), Centriscidae and Macroramphosidae (Centriscoidea), and Syngnathidae and Solenostomidae (Syngnathoidea).

Studies using morphology have proposed monophyly of the order and suggested sister groups based on weak evidence, and while the close relationships of superfamilies is well supported, the relationships among these family pairs remains unclear. In contrast, molecular evidence clearly refutes the monophyly of the Gasterosteiformes, placing gasterosteoids close to the cottoid–zoarcoid lineage (Imamura & Yabe, 2002), excluding the Indostomidae, and placing both groups distant from syngnathoids. While the sister group of syngnathoids remains unknown and the relationships among syngnathoid lineages are poorly resolved, the family Solenostomidae (ghost pipefishes) has been faithfully recovered as the sister group of the Syngnathidae in both morphological and molecular analyses, an evolutionary relationship key to understanding the evolution of male parental care in the family Syngnathidae.

8

Adaptation to Habitat

Fishes that live in surface waters down to about 200 metres, epipelagic fishes, live in a sunlit zone where visual predators use visual systems which are designed pretty much as might be expected. But even so, there can be unusual adaptations. Four-eyed fish have eyes raised above the top of the head and divided in two different parts, so that they can see below and above the water surface at the same time. Four-eyed fish actually have only two eyes, but their eyes are specially adapted for their surface-dwelling lifestyle.

The eyes are positioned on the top of the head, and the fish floats at the water surface with only the lower half of each eye underwater. The two halves are divided by a band of tissue and the eye has two pupils, connected by part of the iris.

The upper half of the eye is adapted for vision in air, the lower half for vision in water. The lens of the eye also changes in thickness top to bottom to account for the difference in the refractive indices of air versus water. These fish spend most of their time at the surface of the water. Their diet mostly consists of the terrestrial insects which are available at the surface.

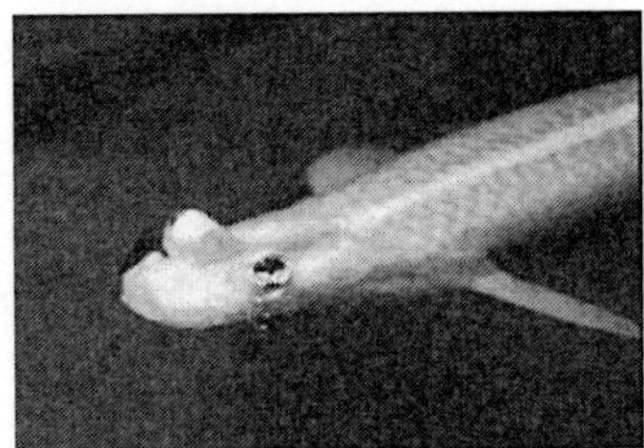

Figure: *The four-eyed fish feeds at the surface of the water with eyes that allow it to see both above and below the surface at the same time*

Figure: *Deepwater fishes, like this Antarctic toothfish, often have large, upward looking eyes, adapted to detect prey silhouetted against the gloom above.*

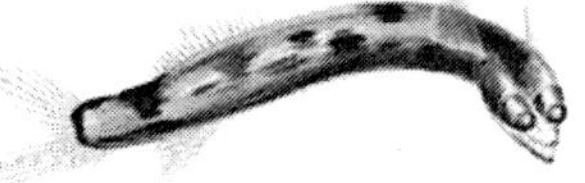

Figure: *The telescopefish has large, forward-pointing telescoping eyes with large lenses.*

Figure: *The mesopelagic sabertooth is an ambush predator with telescopic, upward-pointing eyes.*

Mesopelagic fishes live in deeper waters, in the twilight zone down to depths of 1000 metres, where the amount of sunlight available is not sufficient to support photosynthesis. These fish are adapted for an active life under low light conditions. Most of them are visual predators with large eyes. Some of the deeper water fish have tubular eyes with big lenses and only rod cells that look upwards. These give binocular vision and great sensitivity to small light signals. This adaptation gives improved terminal vision at the expense of lateral vision, and allows the predator to pick out squid, cuttlefish, and smaller fish that are silhouetted against the gloom above them. For more sensitive vision in low light, some fish have a retroreflector behind the retina. Flashlight fish have this plus photophores, which they use in combination to detect eyeshine in other fish.

Still deeper down the water column, below 1000 metres, are found the bathypelagic fishes. At this depth the ocean is pitch black, and the fish are sedentary, adapted to outputting minimum energy in a habitat with very little food and no sunlight. Bioluminescence is the only light available at these depths. This lack of light means the organisms have to rely on senses other than vision. Their eyes are small and may not function at all.

At the very bottom of the ocean flatfish can be found. Flatfish are benthic fish with a negative buoyancy so they can rest on the seafloor. Although flatfish are bottom dwellers, they are not usually deep sea fish, but are found mainly in estuaries and on the continental shelf. When flatfish larvae hatch they have the elongated and symmetric shape of a typical bony fish. The larvae do not dwell on the bottom, but float in the sea as plankton. Eventually they start metamorphosing into the adult form. One of the eyes migrates across the top of the head and onto the other side of the body, leaving the fish blind on one side. The larva loses its swim bladder and spines, and sinks to the bottom, laying its blind side on the underlying surface. Richard Dawkins explains this as an example of evolutionary adaptation

> *...bony fish as a rule have a marked tendency to be flattened in a vertical direction.... It was natural, therefore, that when the ancestors of [flatfish] took to the sea bottom, they should have lain on one* side.... *But this raised the problem that one eye was always looking down into the sand and was effectively useless. In evolution this problem was solved by the lower eye 'moving' round to the upper side.*

Colouration

Fish have evolved sophisticated ways of using colouration. For example, prey fish have ways of using colouration to make it more difficult for visual predators to see them. In pelagic fish, these adaptations are mainly concerned with a reduction in silhouette, a form of camouflage. One method of achieving this is to reduce the area of their shadow by lateral compression of the body. Another method, also a form of camouflage, is by countershading in the case of epipelagic fish and by counter-illumination in the case of mesopelagic fish. Countershading is achieved by colouring the fish with darker pigments at the top and lighter pigments at the bottom in such a way that the colouring matches the background. When seen from the top, the darker dorsal area of the animal blends into the darkness of the water below, and when seen from below, the lighter ventral area blends into the sunlight from the surface. Counter illumination is achieved via bioluminescence by the production of light from ventral photophores, aimed at matching the light intensity from the underside of the fish with the light intensity from the background.

Benthic fish, which rest on the seafloor, physically hide themselves by burrowing into sand or retreating into nooks and crannies, or camouflage themselves by blending into the background or by looking like a rock or piece of seaweed.

While these tools may be effective as predator avoidance mechanisms, they also serve as equally effective tools for the predators themselves. For example, the deepwater velvet belly lantern shark uses counter-illumination to hide from its prey.

The John Dory is a benthopelagic coastal fish with a high laterally compressed body. Its body is so thin that it can hardly be seen from the front. It also has a large dark spot on both sides, which is used to flash an "evil eye" if danger approaches. The large eyes at the front of the head provide it with the bifocal vision and depth perception it needs to catch prey. The John Dory's eye spot on the side of its body also confuses prey, which is then scooped up in its mouth.

Barreleyes

Barreleyes are a family of small, unusual-looking mesopelagic fishes, named for their barrel-shaped, tubular eyes which are generally directed upwards to detect the silhouettes of available prey. Barreleyes have large, telescoping eyes which dominate and protrude from the skull. These eyes generally gaze upwards, but can also be swivelled forwards in some species. Their eyes have a large lens and a retina with an exceptional number of rod cells and a high density of rhodopsin (the "visual purple" pigment); there are no cone cells.

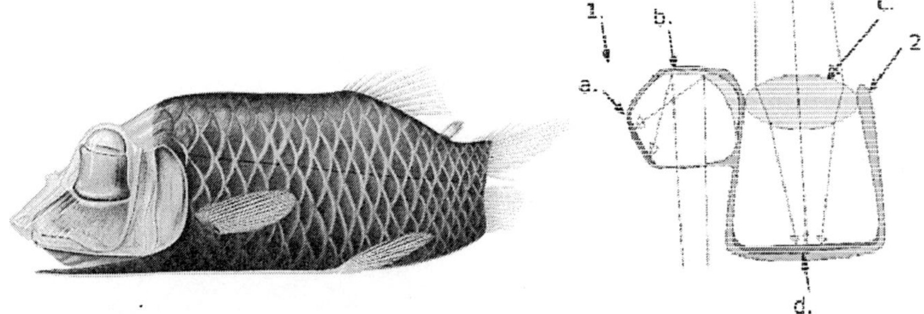

Figure: *Left: The barreleye has barrel-shaped, telescopic eyes which are generally directed upwards, but can also be swivelled forward*
Right: The brownsnout spookfish is the only vertebrate known to employ a mirror eye (as well as a lens): (1) diverticulum (2) main eye (a) retina (b) reflective crystals (c) lens (d) retina

The barreleye species, *Macropinna microstoma*, has a transparent protective dome over the top of its head, somewhat like the dome over an aeroplanee cockpit, through which the lenses of its eyes can be seen. The dome is tough and flexible, and presumably protects the eyes from the nematocysts (stinging cells) of the siphonophores from which it is believed the barreleye steals food.

Another barreleye species, the brownsnout spookfish, is the only vertebrate known to employ a mirror, as opposed to a lens, to focus an image in its eyes. It is unusual in that it utilizes both refractive and reflective optics to see. The main tubular eye contains a lateral ovoid swelling called a diverticulum, largely separated from the eye by a septum.

Other Examples

Small fish often school together for safely. This can have visual advantages, both by visually confusing predator fishes, and by providing many eyes for the school regarded as a body. The "predator confusion effect" is based on the idea that it becomes difficult for predators to pick out individual prey from groups because the many moving targets create a sensory overload of the predator's visual channel. "Shoaling fish are the same size and silvery, so it is difficult for a visually oriented predator to pick an individual out of a mass of twisting, flashing fish and then have enough time to grab its prey before it disappears into the shoal." The "many eyes effect" is based on the idea that as the size of the group increases, the task of scanning the environment for predators can be spread out over many individuals, a mass collaboration presumably providing a higher level of vigilance.

Figure: *The omega iris allows Loricariids to adjust the amount of light that enters their eye*

Fish are normally cold-blooded, with body temperatures the same as the surrounding water. However, some oceanic predatory fish, such as swordfish and some shark and tuna species, can warm parts of their body when they hunt for prey in deep and cold water.

The highly visual swordfish uses a heating system involving its muscles which raises the temperature in its eyes and brain by up to 15 °C. The warming of the retina improves the rate at which the eyes respond to changes in rapid motion made by its prey by as much as ten times.

Many fish have eyeshine. Eyeshine is the result of a light-gathering layer in the eyes called the tapetum lucidum, which reflects white light. It does not occur in humans, but can be seen in other species, such as deer in a headlight.

Eyeshine allows fish to see well in low-light conditions as well as in turbid (stained or rough, breaking) waters, giving them an advantage over their prey. This enhanced vision allows fish to populate the deeper regions in the ocean or a lake. In particular, freshwater walleye are so named because their eyeshine.

Many species of Loricariidae, a family of catfish, have a modified iris called an *omega iris*. The top part of the iris descends to form a loop which can expand and contract called an iris operculum; when light levels are high, the pupil reduces in diametre and the loop expands to cover the centre of the pupil giving rise to a crescent shaped light transmitting portion. This feature gets its name from its similarity to an upside-down Greek letter omega (Ù). The origins of this structure are unknown, but it has been suggested that breaking up the outline of the highly visible eye aids camouflage in what are often highly mottled animals.

Distance Sensory Systems

Visual systems are distance sensory systems which provide fish with data about location or objects at a distance without a need for the fish to directly touch them. Such distance sensing systems are important, because they allow communication with other fish, and provide information about the location of food and predators, and about avoiding obstacles or maintaining position in fish schools. For example, some schooling species have "schooling marks" on their sides, such as visually prominent stripes which provide reference marks and help adjacent fish judge their relative positions. But the visual system is not the only one that can perform such functions. Schooling fish also have a lateral line running the length of their bodies. This lateral line enables the fish to sense changes in water pressure and turbulence adjacent to its body. Using this information, schooling fish can adjust their distance from adjacent fish if they come too close or stray too far.

The visual system in fish is augmented by other sensing systems with comparable or complimentary functions. Some fish are blind, and must rely entirely on alternate sensing systems. Other senses which can also provide data about location or distant objects include hearing and echolocation, electroreception, magnetoception and chemoreception (smell and taste). For example, catfish have chemoreceptors across their entire bodies, which means they "taste" anything they touch and "smell" any chemicals in the water. "In catfish, gustation plays a primary role in the orientation and location of food". Cartilaginous fish (sharks, stingrays and chimaeras) use magnetoception. They possess special electroreceptors called the *ampullae of Lorenzini* which detect a slight variation in electric potential. These receptors, located along the mouth and nose of the fish, operate according to the principle that a time-varying magnetic field moving through a conductor induces an electric potential across the ends of the conductor. The ampullae may also allow the fish to detect changes in water temperature. As in birds, magnetoception may provide information which help the fish map migration routes.

Lateral Line

The lateral line is a sense organ in aquatic organisms (chiefly fish), used to detect movement and vibration in the surrounding water. Lateral lines are usually visible as faint lines running lengthwise down each side, from the vicinity of the gill covers to the base of the tail. Sometimes parts of the lateral organ are modified into electroreceptors, which are organs used to detect electrical impulses. It is possible that vertebrates such as sharks use the lateral organs to detect magnetic fields as well (along with Ampullae of Lorenzini). Most amphibian larvae and some adult amphibians also have a lateral organ. Some crustaceans and cephalopods have similar organs.

Figure: *The lateral line sensory organ, in this case shown on a shark*

Function

The lateral line system allows the detection of movement and vibrations in the water surrounding an animal, providing spatial awareness and the ability to navigate in space. This plays an essential role in orientation, predatory behavior, and social schooling.

In a 2001 study, researchers demonstrated that the lateral line system was necessary to detect vibrations made by prey and to orient towards the source to begin predatory action.

Fish were able to detect movement, produced either by prey or a vibrating metal sphere, and orient themselves toward the source before proceeding to make a predatory strike at it.

This behavior persisted even in blinded fish, but was greatly diminished when lateral line function was inhibited by $CoCl_2$ application.

This cobalt chloride treatment results in the release of Co ions, disrupting ionic transport and preventing signal transduction in the lateral lines.

Further trials utilizing either a gentamicin dip or external scraping of the lateral lines, to disrupt canal and superficial receptors respectively, demonstrated that these behaviors were dependent specifically on mechanoreceptors located within the canals of the lateral line.

Figure: *The small holes on the head of this Northern pike (Esox lucius) contain neuromasts of the lateral line system.*

The role mechanoreception plays in schooling behavior was demonstrated in a 1976 study by Pitcher, et al. A school of Pollachius virens was established in a tank and individual fish were removed and subjected to different procedures before their ability to rejoin the school was observed. Fish that were experimentally blinded were able to reintegrate into the school, while fish with severed lateral lines were unable to reintegrate themselves. Therefore, reliance on functional mechanoreception, not vision, is essential for schooling behavior.

Variations

The development of the lateral-line system depends on the organisms's mode of life. For instance, fish that are active swimming types tend to have more neuromasts in canals than they have on their surface, and the line will be farther away from the pectoral fins, which probably reduces the amount of "noise" that is generated by fin motion.

The lateral-line system helps the fish to avoid collisions, to orient itself in relation to water currents, and to locate prey. For instance, the blind, cave-living Mexican tetra have rows of neuromasts on their heads, which appear to be used to precisely locate food without the use of sight; killifish are able to use their lateral line organ to sense the ripples made by insects struggling on the water's surface. Experiments with pollock have shown that the lateral line is also a key enabler for schooling behaviour.

It has also been suggested that the lateral line may give sharks advanced warning of frontal pressure systems and that they use it to avoid severe weather conditions that may result in injury. It was observed that during Hurricane Gabrielle, which struck Florida in 2001, juvenile black tip sharks moved to deeper waters as the storm approached.

Fish Anatomy

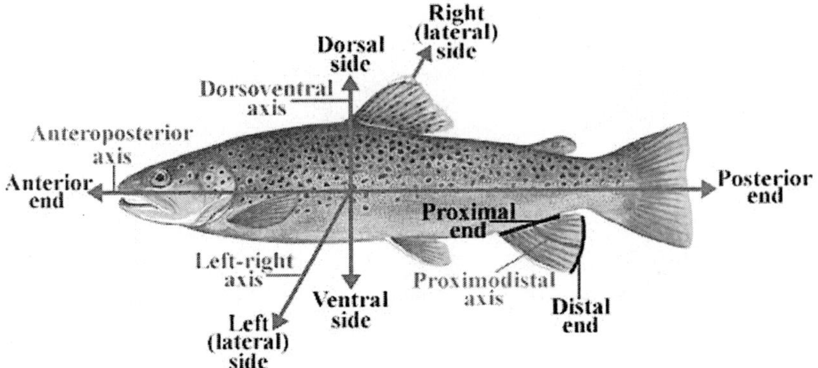

Figure: *Anatomical directions and axes*

Fish anatomy is primarily governed by the physical characteristics of water, which is much denser than air, holds a relatively small amount of dissolved oxygen, and absorbs more light than air does.

Body

Fish have a variety of different body plans. Their body is divided into head, trunk, and tail, although the divisions are not always externally visible. The body is often fusiform, a streamlined body plan often found in fast-moving fish. They may also be filiform (eel-shaped) or vermiform (worm-shaped).

Also, fish are often either laterally (thin) or dorsally (flat) compressed. The caudal peduncle is the narrow part of the fish's body to which the caudal or tail fin is attached. The hypural joint is the joint between the caudal fin and the last of the vertebrae. The hypural is often fan-shaped.

Photophores are light-emitting organs which appears as luminous spots on some fishes. The light can be produced from compounds during the digestion of prey, from specialised mitochondrial cells in the organism called photocytes, or associated with symbiotic bacteria, and are used for attracting food or confusing predators.

The lateral line is a sense organ used to detect movement and vibration in the surrounding water. In most species, it consists of a line of receptors running along each side of the fish. The ampullae of Lorenzini allow sharks to sense electrical discharges. The genital papilla is a small, fleshy tube behind the anus in some fishes, from which the sperm or eggs are released; the sex of a fish often can be determined by the shape of its papilla.

Head

The head includes the snout, from the eye to the forward most point of the upper jaw, the operculum or gill cover (absent in sharks), and the cheek, which extends from eye to preopercle. The operculum and preopercle may or may not have spines. The lower jaw defines a chin. In lampreys, the mouth is formed into an oral disk. In most jawed fish, however, there are three general configurations. The mouth may be on the forward end of the head (terminal), may be upturned (superior), or may be turned downwards or on the bottom of the fish (subterminal or inferior). The mouth may be modified into a suckermouth adapted for clinging onto objects in fast-moving water.

The head may have several fleshy structures known as barbels, which may be very long and resemble whiskers. Many fish species also have a variety of protrusions or spines on the head. The nostrils or nares of almost all fishes do not connect to the oral cavity, but are pits of varying shape and depth.

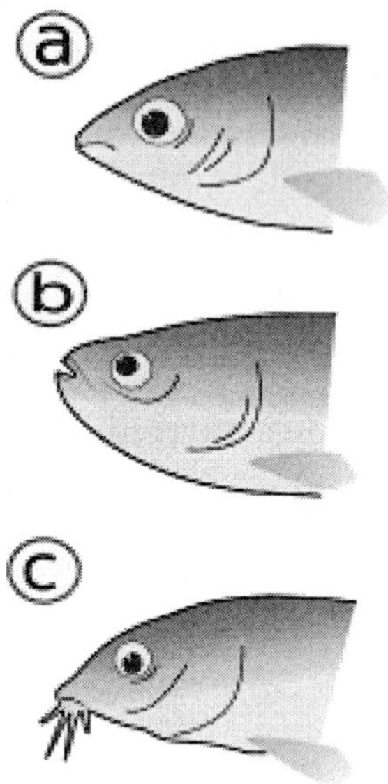

Figure: *Positions of the fish's mouths: (a) terminal, (b) superior, (c) subterminal, inferior*

Fins

The *fins* are the most distinctive features of a fish, composed of bony spines protruding from the body with skin covering them and joining them together, either in a webbed fashion, as seen in most bony fish, or more similar to a flipper, as seen in sharks. These usually serve as a means for the fish to swim. Fins can also be used for gliding or crawling, as seen in the flying fish and frogfish. Fins located in different places on the fish serve different purposes, such as moving forward, turning, and keeping an upright position.

Spines and Rays

In bony fish, most fins may have spines or rays. A fin may contain only spiny rays, only soft rays, or a combination of both. If both are present, the spiny rays are always anterior. Spines are generally stiff and sharp. Rays are generally soft, flexible, segmented, and may be branched. This segmentation of rays is the main difference that separates them from spines; spines may be flexible in certain species, but they will never be segmented. Spines have a variety of uses. In catfish, they are used as a form of defence; many catfish have the ability to lock their spines outwards. Triggerfish also use spines to lock themselves in crevices to prevent them being pulled out.

Lepidotrichia are bony, bilaterally-paired, segmented fin rays found in bony fishes. They develop around actinotrichia as part of the dermal exoskeleton. Lepidotrichia may have some cartilage or bone in them as well. They are actually segmented and appear as a series of disks stacked one on top of another.

Reproductive System

Internal Fertilization: In many species of fish, fins have been modified to allow internal fertilization.

A gonopodium is an anal fin that is modified into an intromittent organ in males of certain species of live-bearing fish in the families Anablepidae and Poeciliidae. It is movable and used to impregnate females during mating. The male's anal fin's 3rd, 4th and 5th rays are formed into a tube like structure in which the sperm of the fish is ejected. In some species, the gonopodium may be as much as 50% of the total body length. Occasionally the fin is too long to be used, as in the "lyretail" breeds of *Xiphophorus helleri*. Hormone treated females may develop gonopodia. These are useless for breeding. One finds similar organs having the same characteristics in other types of fish, for example the andropodium in the *Hemirhamphodon* or in the Goodeidae.

When ready for mating, the gonopodium becomes "erect" and points forward, towards the female. The male shortly inserts the organ into the sex opening of the female, with hook like adaptations that allow the fish to grip onto the female to ensure impregnation. If a female remains stationary and her partner contacts her vent with his gonopodium, she is fertilized. The sperm is preserved in the female's oviduct. This allows females to, at any time, fertilize themselves without further assistance of males. Male cartilaginous fish have claspers modified from pelvic fins. These are intromittent organs, used to channel semen into the female's cloaca during copulation.

Skin

The outer body of many fish is covered with scales. Some species are covered instead by scutes. Others have no outer covering on the skin; these are called naked fish. Most fish are covered in a protective layer of slime (mucus).

There are four types of fish scales.

1. Placoid scales, also called dermal denticles, are similar to teeth in that they are made of dentin covered by enamel. They are typical of sharks and rays.

2. Ganoid scales are flat, basal-looking scales that cover a fish body with little overlapping. They are typical of gar and bichirs.
3. Cycloid scales are small oval-shaped scales with growth rings. Bowfin and remora have cycloid scales.
4. Ctenoid scales are similar to the cycloid scales, with growth rings. They are distinguished by spines that cover one edge. Halibut have this type of scale.

Another, less common, type of scale is the scute, which is:

- an external shield-like bony plate, or
- a modified, thickened scale that often is keeled or spiny, or
- a projecting, modified (rough and strongly ridged) scale, usually associated with the lateral line, or on the caudal peduncle forming caudal keels, or along the ventral profile. Some fish, such as pineconefish, are completely or partially covered in scutes.

Vertebrae

The vertebrae of lobe-finned fishes consist of three discrete bony elements. The vertebral arch surrounds the spinal cord, and is of broadly similar form to that found in most other vertebrates. Just beneath the arch lies a small plate-like pleurocentrum, which protects the upper surface of the notochord, and below that, a larger arch-shaped intercentrum to protect the lower border.

Both of these structures are embedded within a single cylindrical mass of cartilage. A similar arrangement was found in primitive tetrapods, but, in the evolutionary line that led to reptiles (and hence, also to mammals and birds), the intercentrum became partially or wholly replaced by an enlarged pleurocentrum, which in turn became the bony vertebral body.

In most ray-finned fishes, including all teleosts, these two structures are fused with, and embedded within, a solid piece of bone superficially resembling the vertebral body of mammals. In living amphibians, there is simply a cylindrical piece of bone below the vertebral arch, with no trace of the separate elements present in the early tetrapods.

In cartilagenous fish, such as sharks, the vertebrae consist of two cartilagenous tubes. The upper tube is formed from the vertebral arches, but also includes additional cartilaginous structures filling in the gaps between the vertebrae, and so enclosing the spinal cord in an essentially continuous sheath.

The lower tube surrounds the notochord, and has a complex structure, often including multiple layers of calcification.

Lampreys have vertebral arches, but nothing resembling the vertebral bodies found in all higher vertebrates. Even the arches are discontinuous, consisting of separate pieces of arch-shaped cartilage around the spinal cord in most parts of the body, changing to long strips of cartilage above and below in the tail region. Hagfishes lack a true vertebral column, and are therefore not properly considered vertebrates, but a few tiny neural arches are present in the tail.

The Jaw

Linkage systems are widely distributed in animals. The most thorough overview of the different types of linkages in animals has been provided by M. Muller, who also designed a new classification system, which is especially well suited for biological systems.

Linkage mechanisms are especially frequent and manifold in the head of bony fishes, such as wrasses, which have evolved many specialised feeding mechanisms. Especially advanced are the linkage mechanisms of jaw protrusion. For suction feeding a system of linked four-bar linkages is responsible for the coordinated opening of the mouth and 3D expansion of the buccal cavity. Other linkages are responsible for protrusion of the premaxilla.

The vertebrate jaw probably originally evolved in the Silurian period and appeared in the Placoderm fish which further diversified in the Devonian. Jaws are thought to derive from the pharyngeal arches that support the gills in fish. The two most anterior of these arches are thought to have become the jaw itself and the hyoid arch, which braces the jaw against the braincase and increases mechanical efficiency. While there is no fossil evidence directly to support this theory, it makes sense in light of the numbers of pharyngeal arches that are visible in extant jawed (the Gnathostomes), which have seven arches, and primitive jawless vertebrates (the Agnatha), which have nine.

It is thought that the original selective advantage garnered by the jaw was not related to feeding, but to increased respiration efficiency. The jaws were used in the buccal pump (observable in modern fish and amphibians) that pumps water across the gills of fish or air into the lungs in the case of amphibians. Over evolutionary time the more familiar use of jaws (to humans), in feeding, was selected for and became a very important function in vertebrates.

Internal Organs

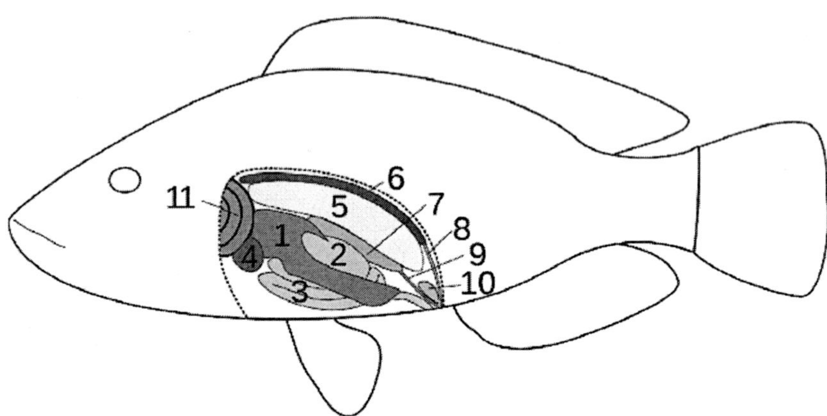

Figure: *Internal Organs*

- The gas bladder, or swim bladder, is an internal organ that contributes to the ability of a fish to control its buoyancy, and thus to stay at the current water depth, ascend, or descend without having to waste energy in swimming. It is often absent in fast swimming fishes such as the tuna and mackerel families.
- Certain groups of fish have modifications to allow them to hear, such as the Weberian apparatus of Ostariophysians.
- The gills, located under the operculum, are a respiratory organ for the extraction of oxygen from water and for the excretion of carbon dioxide. They are not usually visible, but can be seen in some species, such as the frilled shark.
- The labyrinth organ of Anabantoidei and Clariidae is used to allow the fish to extract oxygen from the air.
- Gill rakers are bony or cartilaginous, finger-like projections off the gill arch which function in filter-feeders in retaining prey.
- Electric fish are able to produce electric fields by modified muscles in their body.
- Many fish species are hermaphrodites. *Synchronous hermaphrodites* possess both ovaries and testes at the same time. *Sequential hermaphrodites* have both types of tissue in their gonads, with one type being predominant while the fish belongs to the corresponding gender.
- The blood circulation of fishes is called "single circuit circulatory system."

Electroreception

Electroreception is the biological ability to perceive natural electrical stimuli. It has been observed only in aquatic or amphibious animals, since water is a much better conductor than air. Electroreception is used in electrolocation (detecting objects) and for electrocommunication.

Electroreception is known only in vertebrates. It is found in lampreys, cartilaginous fishes (sharks, rays, chimaeras), lungfishes, bichirs, coelacanths, sturgeons, paddlefishes, catfishes, gymnotiformes, elephantfishes, monotremes, and at least one species of cetacean. The electro receptor organs in all these groups are derived embryologically from a mechanosensory system. In fishes they are developed from the lateral lines. In most groups electroreception is *passive*, where it is used predominantly in predation. Two groups of teleost fishes are weakly electric and engage in *active* electroreception; the Neotropical knifefishes (Gymnotiformes) and the African elephantfishes (Notopteroidei).

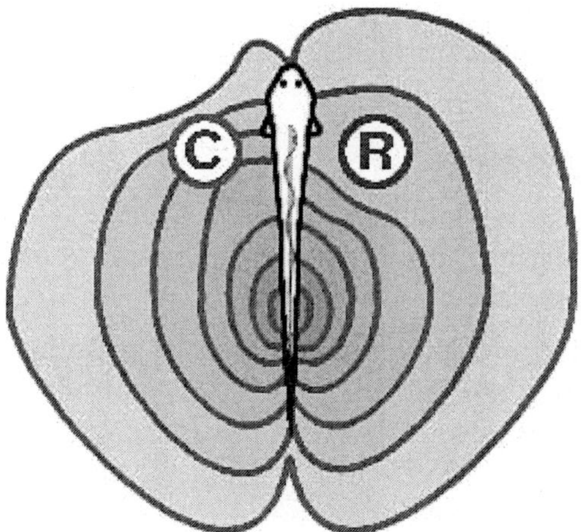

Figure: *Active electrolocation. Conductive objects concentrate the field and resistant objects spread the field.*

Electrolocation

Electroreceptive animals use this sense to locate objects around them. This is important in ecological niches where the animal cannot depend on vision: for example in caves, in murky water and at night. Many fish use electric fields to detect buried prey. Some shark embryos and pups "freeze" when they detect the characteristic electric signal of their predators.

Passive Electrolocation

In passive electrolocation, the animal senses the weak bioelectric fields generated by other animals and uses it to locate them. These electric fields are generated by all animals due to the activity of their nerves and muscles.

A second source of electric fields in fish is the ion pumps associated with osmoregulation at the gill membrane. This field is modulated by the opening and closing of the mouth and gill slits. Many fish that prey on electrogenic fish use the discharges of their prey to detect them. This has driven the prey to evolve more complex or higher frequency signals that are harder to detect.

Passive electroreception is carried out solely by ampullary electroreceptors in fish. It is tuned to low frequency signals (less than 1 Hz to tens of Hz). Fish use passive electroreception to supplement or replace their other senses when detecting prey and predators. In sharks, sensing an electric dipole alone is sufficient to cause them to try and eat it. It has been proposed that sharks can use their acute electric sense to detect the earth's magnetic field by detecting the weak electric currents induced by their swimming or by the flow of ocean currents.

Bioelectromagnetism

Bioelectromagnetism (sometimes equated with bioelectricity) refers to the electrical, magnetic or electromagnetic fields produced by living cells, tissues or organisms.

Examples include the cell membrane potential and the electric currents that flow in nerves and muscles, as a result of action potentials. 'Bioelectromagnetism' is somewhat similar to bioelectromagnetics, which deals with the effect on life from external electromagnetism; yet such an effect also falls under the definition of 'bioelectromagnetism'.

Popular Culture

- Electricity is used to create Frankenstein's monster in the classic story *Frankenstein* as well as most of its adaptations.
- An episode of *Fringe (TV series)*, *Power Hungry*, sees Walter Bishop (John Noble), describe an experiment he once worked on to track humans with pigeons, using their electromagnetic signatures.
- In The Matrix, the machines harness the bioelectricity of humans (along with harvested body heat) to power themselves.

- In the video game, "Deus Ex" and its sequel "Deus Ex: Invisible War", the main characters, JC Denton and Alex D, as well as other characters in the series, utilizes bioelectricity to fuel their biomodifications. It is stored in a metre that depletes with each use of biomodifications, and can be restored with cells or over time throughout the game.
- In the video game, "Yuri's Revenge" expansion of "Red Alert II", Yuri faction's basic power source comes from "Bio-reactors" and garrisoning infantry adds to its total power output.
- Various Sci-Fi shows, such as the Stargate series, include life signs detectors, which could possibly detect bioelectric fields.

Electric Fish

***Figure:** Electric eels are fish capable of generating an electrical field.*

An electric fish is a fish that can generate electric fields. It is said to be *electrogenic*; a fish that has the ability to detect electric fields is said to be *electroreceptive*. Most electrogenic fish are also electroreceptive. Electric fish species can be found both in the sea and in freshwater rivers of South America (Gymnotiformes) and Africa (Mormyridae). Many fish such as sharks, rays and catfishes can detect electric fields, and are thus electroreceptive, but as they cannot generate an electric field they are not classified as electric fish. Most common bony fish (teleosts), including most fish kept in aquaria or caught for food, are neither electrogenic nor electroreceptive.

Aquatic Predation

Aquatic predation presents a special difficulty as compared to predation on land, because the density of water is about the same as that of the prey, so that the prey tends to be pushed away. This problem was first identified by R.N. Alexander. As a result, underwater predators, especially bony fish, have evolved a number of specialised feeding mechanisms, known as filter feeding, ram feeding, suction feeding, protrusion, and pivot feeding.

Most underwater predators combine more than one of these basic principles. For example a typical generalised predator, such as the cod, combines suction with some amount of protrusion and pivot feeding.

Suction Feeding

Suction feeding is a method of ingesting a prey item in fluids by sucking the prey into the predator's mouth. This is typically accomplished by the predator expanding the volume of its oral cavity and/or throat, resulting in a pressure difference between the inside of the mouth and the outside environment. When the mouth is opened, the pressure difference causes water to flow into the predator's mouth, carrying the prey item in with the fluid flow. Bony fish have evolved ingenious mechanisms, so-called mechanical linkages for rapid and forceful expansion of the buccal cavity.

Pivot Feeding

Pivot feeding is a method to transport the mouth towards the prey by an upward turning of the head, which is pivoting on the neck joint. Pipefish such as sea horses and sea dragons are specialised on this feeding mechanism. With prey capture times of down to 5 ms (shrimpfish *Centriscus scutatus*) this method is used by the fastest feeders in the animal kingdom.

The secret of the speed of pivot feeding is in a locking mechanism, in which the hyoid arch is folded under the head and is aligned with the urohyal which connects to the shoulder girdle. The trigger mechanism of unlocking is debated, but is probably in lateral adduction.

Filter Feeding

In filter feeding, the water flow is primarily generated by the organism itself, for example by creating a pressure, by active swimming, or by ciliary movements.

Suspension Feeding

In suspension feeding, the water flow is primarily external or if the particles themselves move with respect to the ambient water, such as in sea lilies.

Bait Ball

A bait ball, or baitball, occurs when small fish swarm in a tightly packed spherical formation about a common centre. It is a last ditch defensive measure adopted by small schooling fish when they are threatened by predators. Small schooling fish are eaten by many types of predators, and for this reason they are called bait fish or forage fish.

For example, sardines group together when they are threatened. This instinctual behaviour is a defence mechanism, as lone individuals are more likely to be eaten than large groups. Sardine bait balls can be 10–20 metres in diametre and extend to a depth of 10 metres. The bait balls are short lived and seldom last longer than 10 minutes.

However, bait balls are also conspicuous, and when schooling fish form a bait ball they can draw the attention of many other predators. As a response to the defensive capabilities of schooling fish, some predators have developed sophisticated countermeasures. These countermeasures can be spectacularly successful, and can seriously undermine the defensive value of forming bait balls.

Predator Cooperation

The most effective strategy predators use against schooling fish is to first scare them into forming a bait ball. Strategies such as those outlined in the previous section, can work to a degree against freely streaming fish schools, but work much better if the fish school is first compacted into a bait ball.

It is difficult for predators working individually to scare a fish school into a bait ball, and they usually work together in a cooperative effort.

- Thresher sharks compact their prey by swimming in circles around them, splashing the water with their long tails, often in pairs or small groups. They then strike sharply at the bait ball to stun the fish with the upper lobe of their tails.
- Schools of forage fish can draw silky sharks in large numbers. Silky sharks have been documented “herding” such schools into a bait ball trapped against the surface, and then consuming the entire school. When attacking tightly packed fish, silky sharks charge through the ball and slash open-mouthed, catching the prey fish at the corners of their jaws. Although multiple individuals may feed at once, each launches its attack independently.

- Pods of many dolphin species commonly herd a school of fish into a bait ball while individual members take turns ploughing through and feeding on the more compacted shoal. *Corralling* is a method where fish are chased to shallow water where they are more easily captured. Some dolphins takes this further with *strand feeding*, driving prey until they are stranded on mud banks, where they can be easily accessed. Dolphins have also been observed blowing bubbles to startle and separate individual fish from a bait ball.
- The humpback whale uses a feeding technique called *bubble net feeding*. A group of whales swim in a shrinking circle blowing bubbles below a school of prey fish. Forage fish show a strong fear of bubbles and can be easily contained within a bubble curtain. The shrinking ring of bubbles encircles the school and confines it in an ever smaller cylinder. The whales then lunge feed, often as a synchronised group, suddenly swimming upward through the "net", mouths agape, swallowing thousands of fish in one gulp. The ring can begin at up to 30 metres (98 ft) in diametre with perhaps a dozen whales cooperating. Using a crittercam attached to whale's back it was discovered that some whales blow the bubbles while others dive deeper to drive fish toward the surface, and yet others herd the prey into the net by vocalising. Some humpback whales also scare schooling fish by slapping their tails (lobtail). Although many whale species lunge feed, only humpbacks use bubble nets.
- Killer whales usually hunt larger fish, such as salmon, individually or in small group of individuals. However, forage fish, such as herrings, are often caught using *carousel feeding*. The killer whales force the herring into a tight ball by releasing bursts of bubbles or flashing their white undersides. They then slap the ball with their tail flukes, either stunning or killing up to 10–15 herring with a successful slap. The herring are then eaten one at a time. Carousel feeding has only been documented in the Norwegian killer whale population and with some oceanic dolphin species.

Cleaner Fish

Cleaner fish are fish that provide a service to other fish species by removing dead skin and ectoparasites. This is an example of mutualism, an ecological interaction that benefits both parties involved.

A wide variety of fishes have been observed to display cleaning behaviors including wrasses, cichlids, catfish, and gobies, as well as by a number of different species of cleaner shrimp. There is also at least one predatory mimic, the sabre-toothed blenny, that mimics cleaner fish but in fact feeds on healthy scales and mucous.

Filter Feeder

Filter feeders (a sub-group of suspension feeders) are animals that feed by straining suspended matter and food particles from water, typically by passing the water over a specialised filtering structure. Some animals that use this method of feeding are clams, krill, sponges, some fish and sharks, and baleen whales. Some birds, such as flamingos, are also filter feeders. Filter feeders can play an important role in clarifying water.

Baleen Whales

The baleen whales, also called whalebone whales or great whales, form the Mysticeti, one of two suborders of the Cetacea (whales, dolphins, and porpoises). Baleen whales are characterised by having baleen plates for filtering food from water, rather than having teeth. This distinguishes them from the other suborder of cetaceans, the toothed whales or Odontoceti. The suborder contains four families and fourteen species. The scientific name derives from the Greek word mystidos, which means "unknowable".

Sponges

Sponges have no true circulatory system; instead, they create a water current which is used for circulation. Dissolved gases are brought to cells and enter the cells via simple diffusion. Metabolic wastes are also transferred to the water through diffusion. Sponges pump remarkable amounts of water. *Leuconia*, for example, is a small leuconoid sponge about 10 cm tall and 1 cm in diametre. It is estimated that water enters through more than 80,000 incurrent canals at a speed of 6 cm per minute. However, because *Leuconia* has more than 2 million flagellated chambers whose combined diametre is much greater than that of the canals, water flow through chambers slows to 3.6 cm per hour. Such a flow rate allows easy food capture by the collar cells.

All water is expelled through a single osculum at a velocity of about 8.5 cm/second: a jet force capable of carrying waste products some distance away from the sponge.

Forage Fish

Forage fish, also called prey fish or bait fish, are small fish which are preyed on by larger predators for food. Predators include other larger fish, seabirds and marine mammals. Typical ocean forage fish feed near the base of the food chain on plankton, often by filter feeding. They include the fishes of the family Clupeidae (herrings, shad, sardines, hilsa, menhaden and sprats), as well as anchovies, capelin and halfbeaks.

Forage fish compensate for their small size by forming schools. Some swim in synchronised grids with their mouths open so they can efficiently filter plankton. These schools can become immense shoals which move along coastlines and migrate across open oceans. The shoals are concentrated fuel resources for the great marine predators. The predators are keenly focused on the shoals, acutely aware of their numbers and whereabouts, and make migrations themselves that can span thousands of miles to connect, or stay connected, with them.

The ocean primary producers, mainly contained in plankton, produce food energy from the sun and are the raw fuel for the ocean food webs. Forage fish transfer this energy by eating the plankton and becoming food themselves for the top predators. In this way, forage fish occupy the central positions in ocean and lake food webs. In recent times, many of the worlds great predator fisheries have collapsed. To compensate, the fishing industry is removing huge amounts of forage fish from the oceans, using factory ships with sophisticated sonar and spotting planes. Most of the catch is fed to farmed animals. Fisheries scientists are expressing concern that this will result in further collapses of the predator fish that depend on them.

In the Oceans

Typical ocean forage fish are small, silvery schooling oily fish such as herring, anchovies and menhaden, and other small, schooling baitfish like capelin, smelts, sand lance, halfbeaks, pollock, butterfish and juvenile rockfish. Herrings are a preeminent forage fish, often marketed as sardines or pilchards.

The term "forage fish" is a term used in fisheries, and is applied also to forage species that are not true fish, but play a significant role as prey for predators. Thus invertebrates such as squid and shrimp are also referred to as "forage fish". Even the tiny shrimp-like creatures called krill, small enough to be eaten by other forage fish, yet large enough to eat the same zooplankton as forage fish, are often classified as "forage fish".

Forage fish utilise the biomass of copepods, mysids and krill in the pelagic zone to become the dominant converters of the enormous ocean production of zooplankton. They are, in turn, central prey items for higher trophic levels. Forage fish may have achieved their dominance because of the way they live in huge, and often extremely fast cruising schools. Though forage fish are abundant, there are relatively few species. There are more species of primary producers and apex predators in the ocean than there are forage fish.

Ocean Food Webs

Forage fish occupy central positions in the ocean food webs. The position that a fish occupies in a food web is called its trophic level (Greek food). The organisms it eats are at a lower trophic level, and the organisms that eat it are at a higher trophic level. Forage fish occupy middle levels in the food web, serving as a dominant prey to higher level fish, seabirds and mammals.

In oceans, most primary production is performed by algae. This is a contrast to land, where most primary production is performed by vascular plants. Algae ranges from single floating cells to attached seaweeds, while vascular plants are represented in the ocean by groups such as the seagrasses. Larger producers, such as seagrasses and seaweeds, are mostly confined to the littoral zone and shallow waters, where they attach to the underlying substrate and still be within the photic zone. Most primary production in the ocean is performed by microscopic organisms, the phytoplankton.

Thus, in ocean environments, the first bottom trophic level is occupied principally by phytoplankton, microscopic drifting organisms, mostly one-celled algae, that float in the sea. Most phytoplankton are too small to be seen individually with the unaided eye. They can appear as a green discolouration of the water when they are present in high enough numbers. Since they increase their biomass mostly through photosynthesis they live in the sun-lit surface layer (euphotic zone) of the sea.

The most important groups of phytoplankton include the diatoms and dinoflagellates. Diatoms are especially important in oceans, where they are estimated to contribute up to 45% of the total ocean's primary production. Diatoms are usually microscopic, although some species can reach up to 2 millimetres in length.

The second trophic level (primary consumers) is occupied by zooplankton which feed off the phytoplankton. Together with the phytoplankton, they form the base of the food pyramid that supports most of the world's great fishing grounds. Zooplankton are tiny animals found with the phytoplankton in oceanic surface waters, and include tiny crustaceans, and fish larvae and fry (recently-hatched fish). Most zooplankton are filter feeders, and they use appendages to strain the phytoplankton in the water.

Some larger zooplankton also feed on smaller zooplankton. Some zooplankton can jump about a bit to avoid predators, but they can't really swim. Like phytoplankton, they float with the currents, tides and winds instead. Zooplanktons can reproduce rapidly, their populations can increase up to thirty percent a day under favourable conditions. Many live short and productive lives and reach maturity quickly.

Particularly important groups of zooplankton are the copepods and krill. These are not shown in the images above, but are discussed in more detail later. Copepods are a group of small crustaceans found in ocean and freshwater habitats. They are the biggest source of protein in the sea, and are important prey for forage fish. Krill constitute the next biggest source of protein. Krill are particularly large predator zooplankton which feed on smaller zooplankton. This means they really belong to the third trophic level, secondary consumers, along with the forage fish.

Together, phytoplankton and zooplankton make up most of the plankton in the sea. Plankton is the term applied to any small drifting organisms that float in the sea (Greek *planktos* = wanderer or drifter). By definition, organisms classified as plankton are unable to swim against ocean currents; they cannot resist the ambient current and control their position. In ocean environments, the first two tropic levels are occupied mainly by plankton. Plankton are divided into producers and consumers. The producers are the phytoplankton (Greek *phyton* = plant) and the consumers, who eat the phytoplankton, are the zooplankton (Greek *zoon* = animal).

Forage Fisheries

Herring has been known as a staple food source since 3000 B.C. In Roman times, anchovies were the base for the fermented fish sauce called *garum*. This staple of cuisine was produced in industrial quantities and transported over long distances.

Fishing for sardela or sardina (*Sardina pilchardus*) is an ongoing activity on the Croatian Adriatic coasts of Dalmatia and Istria. It traces its roots back thousands of years. The region was then largely a Venetian dominion, part of the Roman Empire. The area has always been sustained through fishing mainly sardines. Along the coast towns still promote the traditional practice of fishing by lateen sail boats for tourism and festivals.

Pilchard fishing and processing thrived in Cornwall between 1750 and 1880, after which stocks went into an almost terminal decline. Recently (2007) stocks have been improving. The industry has featured in many works of art, including Stanhope Forbes and other Newlyn School artists.

Contemporary

Traditional commercial fisheries were directed towards high value ocean predators such as cod, rockfish and tuna, rather than humble forage fish. As technologies developed, fisheries became so effective at locating and catching predator fish that many of the stocks collapsed. The industry compensated by turning to species lower in the food chain. In former times, forage fish were more difficult to fish profitably, and were a small part of the global marine fisheries. But modern industrial fishing technologies have enabled the removal of increasing quantities. Industrial-scale forage fish fisheries need large scale landings of fish to return profits. They are dominated by a small number of corporate fishing and processing companies.

Forage fish populations are very vulnerable when faced with modern fishing equipment. They swim near the surface in compacted schools, so they are relatively easy to locate at the surface with sophisticated electronic fishfinders and from above with spotter planes. Once located, they are scooped out of the water using highly efficient nets, such as purse seines, which remove most of the school.

Spawning patterns in forage fish are highly predictable. Some fisheries use knowledge of these patterns to harvest the forage species as they come together to spawn, removing the fish before they have actually spawned. Fishing during spawning periods or at other times when forage fish amass in large numbers can also be a blow to predators.

Many predators, such as whales, tuna and sharks, have evolved to migrate long distances to specific sites for feeding and breeding. Their survival hinges on their finding these forage schools at their feeding grounds. The great ocean predators find that, no matter how they are adapted for speed, size, endurance or stealth, they are on the losing side when faced with the machinery of contemporary industrial fishing.

Altogether, forage fish account for 37 percent (31.5 million tonnes) of all fish taken from the world's oceans each year. However, because there are fewer species of forage fish compared to predator fish, forage species fisheries are the largest in the world. Seven of the top ten fisheries target forage fish. The total world catch of herrings, sardines and anchovies alone in 2005 was 22.4 million tonnes, 24 percent of the total world catch.

The Peruvian anchoveta fishery is now the biggest in the world (10.7 million tonnes in 2004), while the Alaskan pollock fishery in the Bering Sea is the largest single species fishery in the world (3 million tonnes).

The Alaskan pollock is said to be the largest remaining single species source of palatable fish in the world. However, the biomass of pollock has declined in recent years, perhaps spelling trouble for both the Bering Sea ecosystem and the commercial fishery it supports. Acoustic surveys by NOAA indicate that the 2008 pollock population is almost 50 percent lower than last year's survey levels. Some scientists think this decline in Alaska pollock could repeat the collapse experienced by Atlantic cod, which could have negative consequences for the entire Bering Sea ecosystem. Salmon, halibut, endangered Steller sea lions, fur seals, and humpback whales eat pollock and depend on healthy populations to sustain themselves.

Bait and Feeder Fish

Forage fish are sometimes referred to as *bait fish* or *feeder fish.* Bait fish is a term used particularly by recreational fishermen, although commercial fisherman also catch fish to bait longlines and traps. Forage fish is a fisheries term, and is used in the context of fisheries. Bait fish, by contrast, are fish that are caught by humans to use as bait for other fish.

The terms overlap in the sense that most bait fish are also forage fish, and vice versa. Feeder fish is a term used particularly in the context of fish aquariums. It refers essentially to the same concept as forage fish, small fish that are eaten by larger fish, but the term is adapted to the particular requirements of working with fish in aquariums.

Fish Migration

Many types of fish migrate on a regular basis, on time scales ranging from daily to annually or longer, and over distances ranging from a few metres to thousands of kilometres. Fish usually migrate because of diet or reproductive needs, although in some cases the reason for migration remains unknown.

Forage Fish

Forage fish often make great migrations between their spawning, feeding and nursery grounds. Schools of a particular stock usually travel in a triangle between these grounds. For example, one stock of herrings have their spawning ground in southern Norway, their feeding ground in Iceland, and their nursery ground in northern Norway. Wide triangular journeys such as these may be important because forage fish, when feeding, cannot distinguish their own offspring.

Capelin are a forage fish of the smelt family found in the Atlantic and Arctic oceans. In summer, they graze on dense swarms of plankton at the edge of the ice shelf. Larger capelin also eat krill and other crustaceans. The capelin move inshore in large schools to spawn and migrate in spring and summer to feed in plankton rich areas between Iceland, Greenland, and Jan Mayen. The migration is affected by ocean currents. Around Iceland maturing capelin make large northward feeding migrations in spring and summer. The return migration takes place in September to November. The spawning migration starts north of Iceland in December or January.

The diagram on the right shows the main spawning grounds and larval drift routes. Capelin on the way to feeding grounds is coloured green, capelin on the way back is blue, and the breeding grounds are red.

Highly Migratory Species

The term highly migratory species (HMS) has its origins in Article 64 of the United Nations Convention on the Law of the Sea (UNCLOS). The Convention does not provide an operational definition of the term, but in an annex (UNCLOS Annex 1) lists the species considered highly migratory by parties to the Convention. The list includes: tuna and tuna-like species (albacore, bluefin, bigeye tuna, skipjack, yellowfin, blackfin, little tunny, southern bluefin and bullet), pomfret, marlin, sailfish, swordfish, saury and ocean going sharks, dolphins and other cetaceans.

These high trophic oceanodromous species undertake migrations of significant but variable distances across oceans for feeding, often on forage fish, or reproduction, and also have wide geographic distributions.

Thus, these species are found both inside the 200 mile exclusive economic zones and in the high seas outside these zones. They are pelagic species, which means they mostly live in the open ocean and do not live near the sea floor, although they may spend part of their life cycle in nearshore waters.

Highly migratory species can be compared with straddling stock and transboundary stock. Straddling stock range both within an EEZ as well as in the high seas. Transboundary stock range in the EEZs of at least two countries. A stock can be both transboundary and straddling.

Shoaling and Schooling

In biology, any group of fish that stay together for social reasons are said to be shoaling (pronounced), and if, in addition, the group is swimming in the same direction in a coordinated manner, they are said to be schooling (pronounced). In common usage, the terms are sometimes used rather loosely. About one quarter of fishes shoal all their lives, and about one half of fishes shoal for part of their lives.

Fish derive many benefits from shoaling behaviour including defence against predators (through better predator detection and by diluting the chance of individual capture), enhanced foraging success, and higher success in finding a mate. It is also likely that fish benefit from shoal membership through increased hydrodynamic efficiency.

Fish use many traits to choose shoalmates. Generally they prefer larger shoals, shoalmates of their own species, shoalmates similar in size and appearance to themselves, healthy fish, and kin (when recognised).

The “oddity effect” posits that any shoal member that stands out in appearance will be preferentially targeted by predators. This may explain why fish prefer to shoal with individuals that resemble themselves. The oddity effect would thus tend to homogenize shoals.

An aggregation of fish is the general term for any collection of fish that have gathered together in some locality. Fish aggregations can be structured or unstructured. An unstructured aggregation might be a group of mixed species and sizes that have gathered randomly near some local resource, such as food or nesting sites.

If, in addition, the aggregation comes together in an interactive, social way, they are said to be shoaling. Although shoaling fish can relate to each other in a loose way, with each fish swimming and foraging somewhat independently, they are nonetheless aware of the other members of the group as shown by the way they adjust behaviour such as swimming, so as to remains close to the other fish in the group. Shoaling groups can include fish of disparate sizes and can including mixed-species subgroups.

If, as a further addition, the shoal becomes more tightly organised, with the fish synchronising their swimming so they all move at the same speed and in the same direction, then the fish are said to be schooling. Schooling fish are usually of the same species and the same age/size.

Fish schools move with the individual members precisely spaced from each other. The schools undertake complicated manoeuvres, as though the schools have minds of their own.

Shoaling is a special case of aggregating, and schooling is a special case of shoaling. While schooling and shoaling mean different things within biology, they are often treated as synonyms by non-specialists, with speakers of British English tending to use "shoaling" to describe any grouping of fish, while speakers of American English tend to use "schooling" just as loosely.

The intricacies of schooling are far from fully understood, especially the swimming and feeding energetics. Many hypotheses to explain the function of schooling have been suggested, such as better orientation, synchronised hunting, predator confusion and reduced risk of being found.

Schooling also has disadvantages, such as excretion buildup in the breathing media and oxygen and food depletion. The way the fish array in the school probably gives energy saving advantages, though this is controversial.

Fish can be obligate or facultative shoalers. Obligate shoalers, such as tunas, herrings and anchovy, spend all of their time shoaling or schooling, and become agitated if separated from the group. Facultative shoalers, such as Atlantic cod, saiths and some carangids, shoal only some of the time, perhaps for reproductive purposes.

Shoaling fish can shift into a disciplined and coordinated school, then shift back to an amorphous shoal within seconds. Such shifts are triggered by changes of activity from feeding, resting, travelling or avoiding predators.

When schooling fish stop to feed, they break ranks and become shoals. Shoals are more vulnerable to predator attack. The shape a shoal or school takes depends on the type of fish and what the fish are doing. Schools that are travelling can form long thin lines, or squares or ovals or amoeboid shapes. Fast moving schools usually form a wedge shape, while shoals that are feeding tend to become circular.

Forage fish are small fish which are preyed on by larger predators for food. Predators include other larger fish, seabirds and marine mammals. Typical ocean forage fish are small, filter feeding fish such as herring, anchovies and menhaden.

Forage fish compensate for their small size by forming schools. Some swim in synchronised grids with their mouths open so they can efficiently filter feed on plankton.

These schools can become huge, moving along coastlines and migrating across open oceans. The shoals are concentrated fuel resources for the great marine predators.

These sometimes immense gatherings fuel the ocean food web. Most forage fish are pelagic fish, which means they form their schools in open water, and not on or near the bottom (demersal fish). Forage fish are short-lived, and go mostly unnoticed by humans, apart from an occasional support role in a documentary about a great ocean predator.

The predators are keenly focused on the shoals, acutely aware of their numbers and whereabouts, and make migrations themselves, often in schools of their own, that can span thousands of miles to connect with, or stay connected with them.

Herring are among the more spectacular schooling fish. They aggregate together in huge numbers. The largest schools are often formed during migrations by merging with smaller schools.

"Chains" of schools one hundred kilometres long have been observed of mullet migrating in the Caspian Sea. Radakov estimated herring schools in the North Atlantic can occupy up to 4.8 cubic kilometres with fish densities between 0.5 and 1.0 fish/cubic metre. That's about three billion fish in one school.

These schools move along coastlines and traverse the open oceans. Herring schools in general have very precise arrangements which allow the school to maintain relatively constant cruising speeds.

Herrings have excellent hearing, and their schools react very fast to a predator. The herrings keep a certain distance from a moving scuba diver or cruising predator like a killer whale, forming a vacuole which looks like a doughnut from a spotter plane. Many species of large predatory fish also school, including many highly migratory fish, such as tuna and some ocean going sharks. Cetaceans such as dolphins, porpoises and whales, operate in organised social groups called pods.

"Shoaling behaviour is generally described as a trade-off between the anti-predator benefits of living in groups and the costs of increased foraging competition." Landa (1998) argues that the cumulative advantages of shoaling, as elaborated below, are strong selective inducements for fish to join shoals.

Parrish et al. (2002) argue similarly that schooling is a classic example of emergence, where there are properties that are possessed by the school but not by the individual fish. Emergent properties give an evolutionary advantage to members of the school which non members do not receive.

Bibliography

Alexander, Paul: *Sri Lankan Fishermen: Rural Capitalism and Peasant Society*, Sterling, New Delhi, 1982.

Bailey, MT: *A-Z of Tropical Fish Diseases and Health Problems*, Howell Book House, New York, 2001.

Barg, U., and M. J. Phillips: *Environment and Sustainability*, Food and Agriculture Organization of the United Nations, Rome, 1990.

Boserup, E.: *The Conditions of Agricultural Growth: The Economics of Agrarian Change under Population Pressure*, Aldine Press, Chicago, 1965.

Bowman, S.J.: *Digest of Statistics of IAFMM*, Potters Bar, U.K., IAFMM, 1984.

Bull, H. O.: *The Physiology of Fishes*, Academic Press, New York, 1989.

Chow, Ching Kuang: *Fatty Acids in Foods and Their Health Implications*, Routledge Publishing, New York, 2001.

Dholakia, A.D.: *Fisheries and Aquatic Resources of India*, Daya, Delhi, 2004.

Ferguson, H.W.: *Systemic Pathology of Fish*, Iowa State Press, Ames, Iowa, 1989.

Geran, James: *The Proper Care of Goldfish*, TFH Publications, NJ, 2000.

Hamm, Wolf: *Trans Fatty Acids*, Blackwell, NY, 2008.

Hasler, A. D.: *Orientation and Fish Migration, in Hoar, W. S., and Randall, D. J.,* New York, Academic Press, 1971.

Helfman G., Collette B., & Facey D.: *The Diversity of Fishes*, Blackwell Publishing, UK, 2004.

Kulkarni, G.K.: *Fisheries and Fish Toxicology*, APH, Delhi, 2006.

Maddy Hargrove: *The Discus: An Owner's Guide to Happy Healthy Fish*, Howell Books, NY, 1999.

Marshall, N. B.: *The Life of Fishes*, Weidenfield and Nicholson, London, 1965.

Mori, Fumitashi and Sakamoto, Yohei: *Aquarium Fish of the World: The Comprehensive Guide to 650 Species*, Chronicle Books, UK, 1991.

Nash, Colin (2011) *The History of Aquaculture* John Wiley and Sons. ISBN 9780813821634.

Petr, Tomi: *Fisheries in Irrigation Systems of Arid Asia*, Daya, Delhi, 2007.

Ponniah, A.G.: *Fish Biodiversity of India*, NBFGR, 2002.

Ralph, R.: *Breeding Australia's desert goby*, Tropical Fish Hobbyist, 1988.

Rebelin W.E.: *The Pathology of Fishes*, The University of Wisconsin Press, UK, 1975.

Roberts, R.J: *Fish Pathology*, Bailliere Tindall, London, 1989.

Robertson, D.R.: *Fishes of the Tropical Eastern Pacific,* University of Hawaii Press, Honolulu. 1994.

Sahrhage, Dietrich; Johannes Lundbeck: *A History of Fishing. Springer-Verlag,* NY, 1992.

Schmida, Gunther E.: *Rainbow Fish.* Barron's Educational Series, NY, 2000.

Silva, De: *Fish Nutrition in Aquaculture*, Springer, Delhi, 2001.

Singh, Braj Prasad: *Fish Genetic and Biotechnology*, Swastik, Delhi, 2008.

Stoskopf, M.K.: *Fish Medicine*, W.B. Saunders Co. 1993.

Trevor, A. Anderson: *Fish Nutrition in Aquaculture*, Springer, Delhi, 2009.

Tunstall, Jeremy: *The Fisherman: The Sociology of an Extreme Occupation*, McGibbon and Kee, London, 1969.

Untergasser, D.: *Handbook of Fish Diseases*, T.F.H. Publications, Neptune, NJ, 1989.

VanStone, J. W.: *Athapaskan Adaptations: Hunters and Fishermen of the Subarctic Forests*, Aldine, Chicago, 1974.

Ward, Barbara E.: *Varieties of the Conscious Model: Fishermen of South China*, Tavistock, London, 1965.

Ward, Barbara E.: *Varieties of the Conscious Model: Fishermen of South China*, Tavistock, London, 1965.

Warner, William W.: *Beautiful Swimmers: Watermen, Crabs and the Chesapeake Bay*, Little, Brown, Boston, 1976.

Index

A

B

C

D

Distance Sensory Systems, 265.

E

F

❑❑❑